Independent Verification of the Mitigating Systems Performance Index (MSPI) Results for the Pilot Plants

Final Report

U.S. Nuclear Regulatory Commission
Office of Nuclear Regulatory Research
Washington, DC 20555-0001

AVAILABILITY OF REFERENCE MATERIALS
IN NRC PUBLICATIONS

NUREG-1816

Independent Verification of the Mitigating Systems Performance Index (MSPI) Results for the Pilot Plants

Final Report

Manuscript Completed: October 2004
Date Published: February 2005

Prepared by
D.A. Dube [1], C.L. Atwood [2], S.A. Eide [3], B.B. Mrowca [4],
R.W. Youngblood [4], D.P. Zeek [3]

[2] Statwood Consulting
2905 Covington Road
Silver Spring, MD   20910

[3] Idaho National Laboratory
P.O. Box 1625
Idaho Falls, ID 83415-3850

[4] ISL, Inc.
11140 Rockville Pike
Rockville, MD 20852

D.A. Dube, NRC Project Manager

Prepared for
[1] Division of Risk Analysis and Applications
Office of Nuclear Regulatory Research
U.S. Nuclear Regulatory Commission
Washington, DC 20555-0001
NRC Job Codes J8263 and Y6370/Y6636

ABSTRACT

In its Reactor Oversight Process (ROP), the U.S. Nuclear Regulatory Commission (NRC) currently uses performance indicators to quantify safety system unavailability (SSU) for four important nuclear power plant systems. Over time, the NRC staff has identified a number of concerns related to the use of these indicators, including the use of short-term unavailability to approximate unreliability, the use of generic performance thresholds irrespective of variations in risk significance, and potential double-counting as a result of support system failures cascading onto front line systems. Moreover, the way the SSU indicators currently measure unavailability is inconsistent with the definition in the NRC's Maintenance Rule, as well as the indicators promulgated by the World Association of Nuclear Operators and the Institute of Nuclear Power Operations.

This report describes the background, technical issues, and pilot project leading to the development of a more risk-informed performance indicator, known as the Mitigating Systems Performance Index (MSPI). The MSPI addresses most of the concerns related to the use of the current indicators. The NRC staff extensively tested and improved the MSPI methodology during a 12-month pilot plant application phase that involved 20 nuclear power plant units of varying design. The staff also evaluated technical issues related to the new indicator's sensitivity to probabilistic risk assessment (PRA) modeling detail. In addition, the staff compared the MSPI results to the existing indicators, as well as findings from the significance determination process (SDP). The analysis indicates that the MSPI appears to consistently provide a better measure of integrated system performance than the current SSU performance indicators.

ABSTRACT

CONTENTS

Appendices

Figures

Tables

EXECUTIVE SUMMARY

Overview

In its Reactor Oversight Process (ROP), the U.S. Nuclear Regulatory Commission (NRC) currently uses performance indicators to quantify safety system unavailability (SSU) for four important nuclear power plant systems (known as "mitigating systems"). Over time, the NRC staff has identified a number of concerns related to the use of these indicators, including the use of fault exposure hours and short-term unavailability to approximate unreliability, the omission of certain unreliability elements, the use of generic ("one-size-fits-all") performance thresholds irrespective of the risk significance of the system, and potential double-counting as a result of support system failures cascading onto front line systems. Moreover, the way the SSU indicators currently measure unavailability is inconsistent with the definition in the NRC's Maintenance Rule, as well as the indicators promulgated by the World Association of Nuclear Operators and the Institute of Nuclear Power Operations. As a result, nuclear plant system personnel currently have to track plant data in three different ways.

This report describes the background, technical issues, and pilot project leading to the development and independent verification of a more risk-informed performance indicator, known as the Mitigating Systems Performance Index (MSPI). The NRC's Office of Nuclear Regulatory Research (RES) developed the MSPI to address most of the concerns related to the use of the current SSU performance indicators. The purpose of the MSPI is to "monitor the performance of selected systems based on their ability to perform risk-significant functions...." When implemented, the MSPI will replace the existing SSU performance indicators for mitigating systems in the ROP.

The RES staff extensively tested and improved the MSPI methodology during a 12-month pilot plant application phase, initiated in the summer of 2002. The pilot consisted of a 6-month data collection phase by 20 nuclear power plant units of varying design, followed by a 6-month analysis phase. During the pilot program, the RES staff performed the following main activities:

- Verify the reasonableness and accuracy of inputs to the MSPI and results for the 20 plants in the pilot program

- Identify technical issues arising from the formulation of the MSPI, and recommend ways to improve the methodology.

The analysis indicates that the MSPI appears to consistently provide a better measure of integrated system performance than the current SSU performance indicators. The MSPI builds upon the insights and findings developed in the NRC's Risk-Based Performance Indicators (RBPI) program, as described in NUREG-1753, "Risk-Based Performance Indicators: Results of Phase I Development," dated April 2002 (Ref. 1). Toward that end, the MSPI approach separately quantifies the risk significance of changes in unreliability (UR) and unavailability (UA). The approach then rolls these contributions into a single system-level indicator using a calculational algorithm based on Fussell-Vesely (FV) importance measures, thereby avoiding the need for ongoing manipulations of the entire risk model. As currently formulated, the MSPI of a given system is a simplified linear approximation of the change in core damage frequency (CDF) attributable to changes in the reliability and availability of risk-significant elements of the system during internal events with the reactor operating at power.

Results of the Independent Verification

The purpose of the NRC's independent verification of the MSPI was to obtain reasonable assurance of the adequacy of the inputs into the MSPI calculation, and reasonableness of pilot plant results. This was accomplished by assessing the individual inputs to the MSPI calculation on a plant-by-plant, system-by-system, and (in many instances) component-by-component basis. In addition, the verification included comparing the MSPI results using the plant-specific probabilistic risk assessment (PRA) models and standardized plant analysis risk (SPAR) resolution models. Thus, this project included detailed independent verification of the following factors:

- baseline data
- current performance data
- FV/UA and FV/UR importance measures
- electronic spreadsheet calculations
- overall MSPI results

In addition to the independent verification, the RES staff performed analyses to assess the sensitivity of the MSPI results to differences between the licensees' PRA models and the NRC's SPAR models. Finally, the staff compared the MSPI results to the existing SSU performance indicators, as well as findings from the NRC's significance determination process (SDP), as appropriate.

The major findings are as follows:

(1) The generic failure rate values in Table 2 of Appendix F to the draft "Regulatory Assessment Performance Indicator Guideline," which the Nuclear Energy Institute (NEI) promulgated as NEI 99-02 (Ref. 2), are not truly representative of nuclear power plant performance in 1995 – 1997 as supposed, and are not appropriate for use in the MSPI. Consequently, the RES staff has developed an improved set of failure rates. (See Appendix C to this report.)

(2) The independent verification generally showed that the pilot plant submittals for train-specific UA baselines are reasonable. However, the verification identified several baseline UAs that were lower than the unplanned UA values, which is erroneous. Additional guidance and perhaps internal software checks are needed to resolve this discrepancy. Consequently, the RES staff tabulated and compared current UA results across plants and with baselines for the 3-year pilot measurement period, and did not identify any current UA entries as outliers. (See Appendix A to this report.)

(3) The independent verification identified a variety of pilot plant data entry errors, including cases of double- or multiple-counting of failures or demands. The RES staff brought the identified errors to the attention of the licensees, and most were corrected by the time of the final data submittals in March 2003. (See Appendix A to this report.)

(4) The existing SPAR Rev. 3 models had previously been benchmarked against licensee models and, in most cases, were within a factor of 2 – 3 of licensees' PRAs for CDF. However, with regard to risk model importances at the component level, the independent verification identified significant discrepancies between existing SPAR Rev. 3 models and the corresponding plant-specific PRA models. A subsequent effort to enhance the SPAR models succeeded in identifying and resolving many issues related to component FVs. Using the geometric mean (over all monitored components at a plant) as the figure-of-merit, the SPAR resolution models agreed with the 11 unique plant PRA FV/URs within a factor of 2 on average. (See Appendix B to this report.)

(5) The MSPI calculations performed within the NEI spreadsheet were verified by comparing results from an independently developed spreadsheet. Results from both spreadsheets agreed. (See Appendix A to this report.)

(6) Overall, the MSPI results from the pilot plant models were found to be in very good agreement with those of the SPAR resolution models. In terms of color indications, the results from the pilot plant model and SPAR resolution model for the 4th Qtr 2002 are comparable (if not identical), depending on whether the frontstop is used or the effect of common-cause failure modeling is accounted for. Numerical results for MSPI values above the practical limit of significance (1×10^{-7}) generally agreed within a factor of 3. (See Appendix A to this report.)

(7) The detailed analysis of the sensitivity of MSPI results to differences in the licensees' PRA models and the SPAR models demonstrated that these differences should be manageable. For all 11 unique PRA models, only three issues could have a potentially large impact on MSPI results. The study found that significant differences in major model inputs (such as system success criteria or initiating event frequencies) are the primary source of significant quantitative differences, whereas differences on the order of factors of 2 – 3 in basic event probabilities have a much lesser effect on MSPI results. (See Appendix B to this report.)

(8) Recognizing that the methodologies for the MSPI, SDP, and SSU indicators have fundamental differences, the RES staff compared these three measures to determine whether their results showed an overall congruence for all 77 component failures identified during the pilot program. As a result of that exercise, the RES staff concluded that the MSPI is a highly capable performance indicator that can differentiate risk-significant changes in performance and addresses problems associated with the currently used performance indicators. Moreover, the MSPI appears to consistently provide a better measure of integrated system performance than the SSU indicators, while minimizing both false positive and false negative likelihoods to the extent possible. (See Appendix I to this report.)

Major Issues and Recommendations

In the course of the pilot program, a number of significant issues arose regarding the fundamental MSPI methodology, as described in draft NEI 99-02. Resolution of these issues first required a thorough understanding of how each issue affected the MSPI results plant-by-plant, within the group of pilot plants, and across the industry as a whole. These issues relate to:

- the appropriateness of generic baseline reliability data

- "invalid" indicators, whereby one failure beyond normal expectation of performance causes the system to exceed the WHITE threshold

- "insensitive" indicators, whereby a very large number of similar component failures within a system would be necessary to reach the WHITE threshold

- recognition that an increase in unreliability increases the change in CDF (delta CDF or ΔCDF) both through the independent failure contribution and through a common-cause failure contribution

- the concern that because of the prescriptive rules for inclusion of components within the pilot program, some plants may need to monitor an inordinately large number of low risk-significance valves

- the concern that there is inconsistent treatment of support system initiators for safety-related service water and component cooling water from plant-to-plant

On the basis of the issues identified through this study, the staff proposed the following six major recommendations to improve the MSPI, as currently formulated in the draft NEI 99-02:

Recommendation #1: *Table 2 of Appendix F to NEI 99-02 should be revised to use industry failure rates derived for 1999 – 2001 (given in Table C.2 of this report) as a surrogate for 1995 – 1997.*

Recommendation #2: *A "frontstop" (as described in Appendix D to this report) should be used to address the "invalid" indicator issue. The frontstop would take the form of a risk cap of $5x10^{-7}$ on the change in the unreliability index (delta URI) associated with the **single** most risk-significant failure, so long as the delta URI is less than $1x10^{-5}$. The frontstop would only be applied to the GREEN/WHITE threshold.*

Recommendation #3: *The variable "backstop" (as described in Appendix E to this report) should be used to address the "insensitive" indicator issue.*

Recommendation #4: *The MSPI formulation should include the common-cause failure contribution to FV importance (as described in Appendix F to this report), and NEI 99-02 should provide substantial guidance on the process for including this contribution.*

Recommendation #5: *The guidance in Appendix F to NEI 99-02 should be revised to allow licensees the option to exclude low risk valves with Birnbaum importance measures (adjusted for common-cause effects) of less than $1x10^{-6}$/yr (as described in Appendix G to this report).*

Recommendation #6: *The guidance in Appendix F to NEI 99-02 should be revised to require the inclusion of the contribution of cooling water support system initiators to FV importance (as described in Appendix H to this report).*

The NRC staff has not yet resolved all issues identified during the course of the pilot program; however, the above recommendations address the major *technical* issues associated with the proposed MSPI formulation. The staff continues to address other issues, which largely relate to the *implementation* of the MSPI. In addition, the guidance in Appendix F to the draft NEI 99-02 continues to be modified to incorporate findings resulting from this research. Finally, it should be noted that a separate task group is developing a process to identify and resolve potentially significant modeling differences between the licensees' PRA models and the NRC's SPAR models.

FOREWORD

Several years ago, the U.S. Nuclear Regulatory Commission (NRC) revamped its inspection, assessment, and enforcement programs for commercial nuclear power plants. As a result, the new Reactor Oversight Process uses more objective, timely, and safety-significant criteria in assessing performance, while seeking to more effectively and efficiently regulate the industry. In particular, the oversight process focuses on activities that pose the greatest potential risks in nuclear power plant operations. The fundamental principle is that the NRC should focus increased regulatory attention on nuclear power plants that exhibit performance problems, while maintaining a normal level of regulatory attention for facilities that perform well.

Inherent in this new process is the use of objective measures of nuclear power plant performance. The NRC measures plant performance through its inspection program and using a combination of objective performance indicators. These performance indicators use objective data to monitor performance within each of seven "cornerstones of safety," including one that encompasses four important nuclear power plant systems (known as "mitigating systems"). Each performance indicator is measured against thresholds that relate to the given indicator's effect on safety. The NRC staff then reviews the performance indicators and posts them on the agency's public Web site.

Since the inception of the reactor oversight process, ongoing concerns have arisen regarding the performance indicators for the Mitigating Systems cornerstone. In particular, the existing safety system unavailability performance indicators account only for system *unavailability* (that is, the fraction of time during plant operation during which equipment in the system is down for repair and maintenance). Moreover, the way the safety system unavailability indicators currently measure unavailability is inconsistent with the definition in the NRC's Maintenance Rule, and the indicators issued by the World Association of Nuclear Operators and the Institute of Nuclear Power Operations. As a result, nuclear plant system personnel currently have to track plant data in three different ways. The safety system unavailability indicators also do not measure unreliability (that is, the likelihood that the equipment would fail to perform during a serious event or emergency). In addition, the current thresholds for action in the safety system unavailability indicators apply (for the most part) generically across all plants, and do not account for the large dissimilarities in design and operation of the numerous (more than 100) nuclear power plant units in the country.

To address these concerns, the industry proposed a revision to the "Regulatory Assessment Performance Indicator Guideline," which the Nuclear Energy Institute published as NEI 99-02. The new approach, known as the Mitigating Systems Performance Index, would measure safety system performance by addressing both unavailability and unreliability. However, the new approach would depart from existing indicators in that it would assign the greatest weight to the most risk-significant equipment in each of six systems at a particular plant. Consequently, the NRC's Office of Nuclear Regulatory Research initiated a 12-month pilot program, consisting of 6 months of data collection and 6 months of analysis to assess the proposed approach.

The primary goal of the research described in this report was to independently verify the results of the pilot program. The research was conducted by NRC staff in the Office of Nuclear Regulatory Research, Division of Risk Analysis and Applications. The two principal contractors were Idaho National Laboratory and Information Systems Laboratories. The research was performed by obtaining the monthly plant performance data and results from the 20 pilot plant submittals over a period of 6 months. The NRC and its contractors then verified the reasonableness of the data, and the NRC used its own risk assessment models and plant equipment performance data to provide an independent assessment of the results. The NRC and its contractors also performed many sensitivity studies to determine how differences in plant data and risk assessment models could affect the overall results. In addition, each recommendation that the staff has proposed in this report was extensively tested by assessing the effect of the recommendation on the indicator results through direct calculation and/or numerical simulation.

On the basis of this research, the staff concludes that the Mitigating Systems Performance Index, as modified by the NRC staff, is a highly capable performance indicator that can differentiate risk-significant changes in performance. The modified index also addresses problems associated with the currently used safety system unavailability performance indicators. As a result, the Mitigating Systems Performance Index appears to consistently provide a better overall measure of integrated system performance than the current safety system unavailability indicators.

When implemented, the Mitigating Systems Performance Index will result in a change to the way that reactor safety system performance is measured and reported on the NRC's public Web site. Moreover, the recommendations of this report will result in an improvement to the method that was originally proposed in the draft of NEI 99-02.

Carl J. Paperiello, Director
Office of Nuclear Regulatory Research
U.S. Nuclear Regulatory Commission

ACKNOWLEDGMENTS

The authors acknowledge the support and collaboration provided by our many colleagues at the U.S. Nuclear Regulatory Commission (NRC), Idaho National Laboratory, and Information Systems Laboratories (ISL). In particular, Patrick Baranowsky of the NRC contributed substantially to many of the conceptual issues and technical ideas, and supported the overall project since its inception. James Houghton and Marcel Harper of the NRC assisted in the review and compilation of findings from the NRC's significance determination process and safety system unavailability indicators. Christopher Hunter of the NRC prepared the appendix on public comments and the related NRC responses. In addition, the research benefited from important discussions during the numerous public meetings concerning the Mitigating Systems Performance Index, as well as meetings with the NRC's Advisory Committee on Reactor Safeguards. In addition, the authors are especially grateful to Colleen Amoruso and Jennifer Jones of ISL, and Paula Garrity of the NRC, for production of this report.

Acknowledgments

ABBREVIATIONS

AC	alternating current
ACP	alternating current power
ACRS	Advisory Committee on Reactor Safeguards
AFW	auxiliary feedwater (system)
AOT	allowed outage time
AOV	air-operated valve
B	Birnbaum Importance
B&W	Babcock & Wilcox
BWR	boiling-water reactor
CCF	common-cause failure
CCW	component cooling water (system)
CDE	Consolidated Data Entry program
CDF	core damage frequency
CNIP	constrained non-informative prior
CVCS	chemical and volume control system
DC	direct current
DCP	direct current power
DDP	diesel-driven pump
DG	diesel generator
DRAA	Division of Risk Analysis and Applications
EAC	emergency alternating current (power system)
EB	empirical Bayes
EDG	emergency diesel generator
EPIX	Equipment Performance and Information Exchange
FTLR	failure to load and run
FTO/C	failure to open or close
FTR	failure to run
FTS	failure to start
FV	Fussell-Vesely Importance
HPCI	high-pressure coolant injection (system)
HPCS	high-pressure core spray (system)
HPI	high-pressure injection (may include other high-head systems, such as CVCS)
HPSI	high-pressure safety injection (system)
HRS	heat removal system
IC	isolation condenser
ICCDP	incremental conditional core damage probability
INPO	Institute for Nuclear Power Operations
LER	licensee event report
LOCA	loss-of-coolant accident
LOOP	loss of offsite power

MDP	motor-driven pump
MLE	maximum likelihood estimate
MOV	motor-operated valve
MSPI	Mitigating System Performance Index
NEI	Nuclear Energy Institute
NRC	U.S. Nuclear Regulatory Commission
NRR	Office of Nuclear Reactor Regulation (NRC)
PCS	power conversion system
PM	preventive maintenance
PORV	power-operated relief valve
PRA	probabilistic risk assessment
RADS	Reliability and Availability Database System
RAW	risk achievement worth
RBPI	risk-based performance indicator
RC	risk cap
RCIC	reactor core isolation cooling (system)
RCP	reactor coolant pump
RES	Office of Nuclear Regulatory Research (NRC)
RHR	residual heat removal (system)
ROP	Reactor Oversight Process
SBO	station blackout
SDP	Significance Determination Process
SGTR	steam generator tube rupture
SPAR	Standardized Plant Analysis Risk
SSU	safety system unavailability
SWS	service water system
TDP	turbine-driven pump
TI	Temporary Instruction
TS	Technical Specification(s)
UA	unavailability
UAI	unavailability index
UR	unreliability
URI	unreliability index

1. INTRODUCTION

1.1 Background

In its Reactor Oversight Process (ROP), the U.S. Nuclear Regulatory Commission (NRC) currently uses performance indicators to quantify safety system unavailability (SSU) for four important nuclear power plant systems (known as "mitigating systems"). Over time, the NRC staff has identified a number of concerns related to the use of these indicators, including the use of fault exposure hours and short-term unavailability to approximate unreliability, the omission of certain unreliability elements, the use of generic ("one-size-fits-all") performance thresholds irrespective of the risk significance of the system, and potential double-counting as a result of support system failures cascading onto front line systems. Moreover, the way the SSU indicators currently measure unavailability is inconsistent with the definition in the NRC's Maintenance Rule, as well as the indicators promulgated by the World Association of Nuclear Operators and the Institute of Nuclear Power Operations. As a result, nuclear plant system personnel currently have to track plant data in three different ways.

In an effort to address these concerns, the NRC initiated the Risk-Based Performance Indicator (RBPI) development program. In Phase I of that program, as described in NUREG-1753, "Risk-Based Performance Indicators: Results of Phase I Development," dated April 2002 (Ref. 1), the NRC explored several possible enhancements to the ROP performance indicators. A key aspect of the approach discussed in Ref. 1 was the use of plant-specific standardized plant analysis risk (SPAR) models to assess the risk significance of changes in unreliability (UR) and unavailability (UA). Based on these models, it was possible to develop candidate RBPIs that separately quantify UR and UA within a common model framework. It was also possible to determine plant-specific thresholds for these indicators. These enhancements help to address the issues associated with the current ROP indicators. In Phase 1 of the RBPI program, the NRC demonstrated that these enhancements are generally feasible, although statistical uncertainty is an issue for some UR indicators.

Thus, although these candidate RBPIs display certain benefits compared to the SSU performance indicators that are currently in use, they also have certain drawbacks. In particular, implementing separate train-level UR and UA indicators leads to a substantial increase in the number of indicators. This increase would raise concerns regarding their effect on the action matrix in the ROP, if implemented. In addition, a larger number of indicators increases the likelihood that at least one indicator will give a false indication.

1.2 Purpose and Scope

This report describes the background, technical issues, and pilot project leading to the development and independent verification of a more risk-informed performance indicator, known as the Mitigating Systems Performance Index (MSPI). The NRC's Office of Nuclear Regulatory Research (RES) developed the MSPI to address most of the concerns related to the use of the current SSU performance indicators. The purpose of the MSPI is to "monitor the performance of selected systems based on their ability to perform risk-significant functions…." When implemented, the MSPI will replace the existing SSU performance indicators for mitigating systems in the ROP.

The RES staff extensively tested and improved the MSPI methodology during a 12-month pilot plant application phase, initiated in the summer of 2002. The pilot consisted of a 6-month data collection phase by 20 nuclear power plant units of varying design (from September 2002 through February 2003), followed by a 6-month analysis phase. During the pilot program, the RES staff performed the following main activities:

- Verify the reasonableness and accuracy of inputs to the MSPI and results for the 20 plants in the pilot program

- Identify technical issues arising from the formulation of the MSPI, and recommend ways to improve the methodology.

The purpose of the NRC's independent verification of the MSPI was to obtain reasonable assurance of the adequacy of the inputs into the MSPI calculation, and reasonableness of pilot plant results. This was accomplished by assessing the individual inputs to the MSPI calculation on a plant-by-plant, system-by-system, and (in many instances) component-by-component basis. In addition, the verification included comparing the MSPI results using the plant-specific probabilistic risk assessment (PRA) models and SPAR resolution models. Thus, this project included detailed independent verification of the following factors:

- baseline data
- current performance data
- FV/UA and FV/UR importance measures
- electronic spreadsheet calculations
- overall MSPI results

In addition to the independent verification, the RES staff performed analyses to assess the sensitivity of the MSPI results to differences between the licensees' PRA models and the NRC's SPAR models. Finally, the staff compared the MSPI results to the existing SSU performance indicators, as well as findings from the NRC's significance determination process (SDP), as appropriate.

The analysis indicates that the MSPI appears to consistently provide a better measure of integrated system performance than the current SSU performance indicators. The MSPI builds upon the insights and findings developed in the NRC's RBPI development program, as described in NUREG-1753 (Ref. 1). Toward that end, the MSPI approach separately quantifies the risk significance of changes in unreliability (UR) and unavailability (UA). The approach then rolls these contributions into a single system-level indicator using a calculational algorithm based on Fussell-Vesely importance measures, thereby avoiding the need for ongoing manipulations of the entire risk model. As currently formulated, the MSPI of a given system is a simplified linear approximation of the change in core damage frequency (CDF) attributable to changes in the reliability and availability of risk-significant elements of the system during internal events with the reactor operating at power. This approach is quantitatively adequate until changes in UR and UA become very large, at which point the numerical inaccuracy can be considerable. However, licensee and regulatory attention would have become focused on these contributions by then. A full discussion of the limitations of the linearized approximation is provided in Appendix M.

The body of this report provides an overview of the RES findings and results, while the appendices augment that overview with technical details. Program and implementation issues associated with the MSPI are beyond the scope of this report and will be addressed in separate assessments in conjunction with the NRC's Office of Nuclear Reactor Regulation (NRR).

2. CHARACTERIZATION OF MSPIs

2.1 Purpose of MSPIs

The staff of the NRC's Office of Nuclear Regulatory Research (RES) developed the Mitigating Systems Performance Index (MSPI) to address concerns (discussed in Section 1.1) related to the use of the current safety system unavailability (SSU) performance indicators in the NRC's Reactor Oversight Process (ROP). According to the draft "Regulatory Assessment Performance Indicator Guideline," which the Nuclear Energy Institute (NEI) promulgated as NEI 99-02 (Ref. 2), the purpose of the MSPI is to "monitor the performance of selected systems based on their ability to perform risk-significant functions...." As such, the new MSPI approach measures safety system performance by addressing both unavailability and unreliability. When implemented, the MSPI will replace the existing SSU performance indicators for mitigating systems in the ROP.

2.2 Definition of MSPIs

In the current NEI formulation, the MSPI of a given system is a simplified linear approximation of the change in core damage frequency (CDF) attributable to changes in the reliability and availability of risk-significant elements of the system during internal events with the reactor operating at power. Thus, the calculation focuses on key components, and quantifies the change in CDF using a simple formula based on the sum of changes in the unavailability index (UAI) and the unreliability index (URI), as follows:

$$MSPI = UAI + URI$$

The Unavailability-Related Contribution

The contribution related to unavailability (UA), is a sum of UA contributions from different trains:

$$UAI = \sum_{j=1}^{n} UAI_{tj} \qquad \text{(Eq. 1)}$$

The summation runs across trains, and UAI_{tj} is the contribution of the jth train to the change in CDF attributable to changes in the UA of that train, according to the formulation in NEI 99-02.

If contributions to the UA of a given train can be collected into a single PRA basic event having unavailability UA_t, the change in CDF associated with a change in train UA can be written as follows (Ref. 2):

$$UAI_t = B(UA) * \Delta UA$$
$$UAI_t = B(UA) * (UA_t - UA_{BLt})$$
$$UAI_t = CDF_p \left[\frac{FV_{UAp}}{UA_p} \right] (UA_t - UA_{BLt}),$$

where $B(UA)$ is the Birnbaum importance for UA, FV_{UA} is the Fussell-Vesely importance for UA, UA_{BLt} is the baseline unavailability, and ΔUA is the change (or "delta") in unavailability.

Items carrying a "p" subscript are understood to be calculated using the PRA values, while those on the right-hand side carrying a "t" subscript (referring to "train" in the NEI formulation) instead of a "p" subscript are derived from either current operating data or baseline data. This formulation divorces the calculation of B(UA) from the calculation of ΔUA. In other words, B(UA) is independent of the value of UA. Given B(UA), the terms for which the difference yields ΔUA need only be calculated on a mutually consistent basis — not necessarily consistently with the PRA — in order for the formula to yield a good estimate of ΔCDF. Of course, if CDF and FV are calculated and combined as above, CDF and FV both need to be based on the same value of UA that appears in the denominator in order to yield B(UA) as desired.

In practice, UA data are collected on a train basis. This avoids the potential overestimation of train unavailability that could result if the unavailabilities of individual components were collected and summed as if they were independent. If one has separate terms in Equation 1 that cannot be collected into a single basic event, each element of the sum can still be calculated using the above approach.

The Unreliability-Related Contribution
The treatment of the UR-related contribution generally follows the above treatment of UAI. However, the elemental contributions to train unreliability must be assessed separately, and partly as a result of this, there are additional considerations for URI.

The following formula is used for the UR-related contribution:

$$URI = CDF_P \sum_{j=1}^{m} \left[\frac{FV_{URcj}}{UR_{pcj}} \right]_{max} (UR_{Bcj} - UR_{BLcj}) ,$$

where

the summation is over those active components and failure modes in the system that can by themselves fail a "train,"

CDF_p is the plant-specific core damage frequency for internal events at power,

FV_{URcj} is the Fussell-Vesely value for unreliability of component j,

UR_{pcj} is the plant-specific PRA value for unreliability of component j,

UR_{Bcj} is the current estimate of ("Bayesian corrected") unreliability of component j for the previous 12 quarters,

UR_{BLcj} is the historical baseline unreliability for component j, and

Max refers to using the highest FV/UR from all the basic events (i.e., failure modes) for a given component (see Ref. 2).

Note that the current formulation considers only the internal events initiators, and **does not** include internal flooding events or external events initiators.

The current estimate is the Bayes posterior mean, based on the most recent 12 quarters of data and the constrained non-informative prior (CNIP). Appendix J presents the technical basis for the use of the CNIP. Because the unreliability parameters are considered variable over time, data are not accumulated over more than 12 quarters; instead, the formulation uses the same CNIP as the prior with each new 12-quarter data set.

Given the above formulations of UAI and URI, a simple qualitative definition is that the MSPI is *a measure of the deviation of actual plant system unavailability and component unreliabilities from baseline values, weighted by plant-specific risk importance measures.*

Effect of Linear Approximation

The MSPI is a linear approximation of changes in CDF attributable to changes in UR and UA. Consider the leading terms in the Taylor expansion of CDF as a function of two changing quantities A and C (which could be two URs, two UAs, or one of each):

$$CDF(A_0 + \Delta A, C_0 + \Delta C) \approx CDF(A_0, C_0) +$$

$$\left.\frac{\partial CDF}{\partial A}\right|_{0,0} \Delta A + \left.\frac{\partial CDF}{\partial C}\right|_{0,0} \Delta C +$$

$$\frac{1}{2} \left.\frac{\partial^2 CDF}{\partial A^2}\right|_{0,0} (\Delta A)^2 + \left.\frac{\partial^2 CDF}{\partial A \partial C}\right|_{0,0} \Delta A \Delta C + \frac{1}{2} \left.\frac{\partial^2 CDF}{\partial C^2}\right|_{0,0} (\Delta C)^2 + ...$$

The MSPI approximation of the change in CDF corresponds to the first-derivative terms in this expansion; the MSPI currently neglects the second- and higher-order corrections. This MSPI approximation is good for small changes in UR or UA; however, for larger changes, the MSPI underestimates the change in CDF, if multiple changing quantities appear together in cut sets. Appendix M presents a quantitative analysis of this underestimation.

2.3   Benefits of MSPIs

Two key attributes of the MSPI are the consistent treatment of both unavailability and unreliability, and the implementation of plant-specific performance thresholds. Basic reliability theory recognizes that optimum system performance is achieved when the proper amount of preventive maintenance (PM) is applied. (Too little PM causes the unreliability term to become unacceptably high, while too much PM drives the unreliability term to near-zero but at the expense of too much downtime.) In addition, the implementation of plant-specific thresholds acknowledges the large dissimilarities in design and operation of nuclear power plants, and sets the performance thresholds commensurate with the risk-significance of the varying systems and the number of component demands and failures.

More specifically, the MSPI offers the following significant benefits:

(1) The MSPI treats UR as it is treated in NUREG-1753. This treatment is based on failure and demand counts, rather than fault exposure time. The MSPI is intended to resolve certain issues associated with the way in which the existing SSU performance indicators treat fault exposure time, including the tendency of the T/2 fault exposure hours (i.e. one-half of the test interval) to overestimate risk significance in the current SSU indicators.

(2) The MSPI is simple to calculate *in comparison to complete PRA model requantification.* An MSPI requires only baseline performance parameters that go into the prior distributions for Bayesian updating, and a set of importance measures derived from a plant model. Once the importance measures are derived, it is no longer necessary to manipulate the plant model in order to quantify the MSPI. Instead, given the above parameters, the MSPI can be quantified by hand calculation (although a spreadsheet, database, or computer program is normally preferred).

(3) The MSPI rolls up most equipment performance data into a single performance-related figure of merit for each system. Therefore, although the MSPI addresses both reliability and availability, and although it spans non-diverse trains within a given system, the number of different performance indices that would result from the various combinations is kept to a minimum.

(4) The MSPI directly measures the performance of cooling water support systems (e.g., service water and component cooling water). As a result, there is no longer a need for cooling water support system failures to "cascade" onto mitigating systems as is the current practice with the SSU indicators.

(5) The MSPI can be a very good approximation to the change in CDF attributable to current performance, provided that changes in performance are not extremely large, and provided that current performance can be accurately estimated. If the changes in performance are large, the correspondence between the MSPI and the change in CDF loses numerical accuracy, but the MSPI still points to the existence of a large change.

2.4   Limitations of MSPIs

This section discusses some of the MSPI limitations denoted in NEI 99-02. First, because of the limitations of the index, the following conditions rely upon the inspection process to evaluate performance issues:

(1) multiple concurrent failures of components, including common-cause failures

(2) conditions that cannot be discovered during normal surveillance tests

(3) failures of inactive components (such as piping and heat exchangers) that are not accounted for in train unavailability

Based upon the pilot program results, the treatment of these conditions under the inspection process is reasonable, and the NRC staff does not recommend any deletions from or additions to the above list. Nonetheless, the possibility of having so-called "invalid indicators" defer to the inspection process was discussed during the pilot program. "Invalid indicators" are MSPIs that have the property that just one failure above baseline during the observation period can cause the index to go "WHITE." Appendix F to NEI 99-02 (Ref. 2) states this property as follows:

> If, for any failure mode for any component in a system, the risk increase (ΔCDF) associated with the change in unreliability resulting from single failure is larger than 1.0×10^{-6}, then the performance index will be considered invalid for that system.

If implemented as proposed in Section 5.2 and Appendix D to this report, the recommended "frontstop" concept could obviate the need to have invalid indicators defer exclusively to the inspection process as originally proposed.

Concerns have also arisen regarding the seeming insensitivity of some indicators. Specifically, in some systems at some plants, a very large number of failures is needed to yield a "WHITE" in the MSPI as currently formulated. Thus, on a few occasions, the MSPI seems hypersensitive, while on a few other occasions, it seems insensitive. Preliminary work has been done on an approach involving the use of a "mixture prior," which might help with both sensitivity extremes. (Appendices D and J to this report briefly discuss mixture priors.) Using a mixture prior can yield a decision rule that, in effect, shows when a frontstop is justified on the basis of prior belief and current evidence. Moreover, updating the same mixture prior with a significantly *larger* number of failures yields an estimate of ΔUR that can be *greater than* the estimate from the CNIP.

However, the mixture prior requires specification of more parameters than would be required for the CNIP, and these values have not yet been determined. For this and other reasons more work would be needed before a mixture prior could replace the CNIP in the MSPI. However, this enhancement to the MSPI could significantly reduce the need to consider post-processing steps such as the frontstop. Once the additional parameters are determined and entered into the MSPI algorithm, no additional effort would be required during implementation.

Finally, as mentioned in Section 2.2 (above), the MSPI can underestimate ΔCDF under some conditions. Specifically, this occurs when significant changes arise in the URI or UAI of elements that appear together in cut sets. It presently appears that, most of the time, this so-called "cohort effect" will not seriously affect the MSPI estimate, partly because significant changes in multiple items are infrequent. However, there is some potential for a significant cohort effect to occur. (Appendix M to this report discusses the cohort effect in greater detail.) It may be possible to compensate for this effect with a simple correction of the MSPI formula, reflecting higher-order contributions to ΔCDF, as mentioned in Section 2.2. Once the parameters needed to quantify higher-order corrections are determined and entered into the MSPI algorithm, no additional effort would be required during implementation; however, numerous parameters will need to be determined.

Despite its benefits, the MSPI also has a number of drawbacks. These include much greater licensee effort (compared to the current SSU performance indicators) in terms of up-front identification of system boundaries and components, assembly and adjustment (if necessary) of component risk measures, and data tabulation. Furthermore, quantification of the MSPI is not as transparent as it is for the current SSU indicators. Discussion and resolution of implementation issues are beyond the scope of this report and will be addressed in separate assessments in conjunction with the NRC's Office of Nuclear Reactor Regulation (NRR).

3.   DESCRIPTION OF THE PILOT PROGRAM

3.1   Objectives and Participants

At the onset of the pilot, the staff had three primary objectives:

(1) Exercise the MSPI guidance.
(2) Perform validation and verification.
(3) Perform Temporary Instruction inspections.

The first objective was primarily addressed by the licensees of the 20 nuclear power plant units that participated in the pilot. With Ref. 2 as guidance, this objective entailed the following activities:

- Identify risk-significant functions for the six systems of interest.

- Identify success criteria.

- Identify data sources.

- Identify system boundaries.

- Identify active components to be monitored.

- Tabulate Fussell-Vesely importance measures and basic event probabilities for all components to be monitored.

- Collect relevant unavailability and unreliability data.

- Populate the pre-formatted NEI electronic worksheets.

- Compute UAI, URI, and MSPI results on a quarterly basis, and submit ongoing results to the NRC on a monthly basis (for this pilot only).

- Identify possible "invalid indicators."

- Assess the reasonableness of results.

The RES staff assumed responsibility for the second objective, which involved a plant-by-plant cross-comparison of performance data, use of SPAR models to validate importance measures, and identification and resolution of significant issues concerning the MSPI methodology. In order to reconcile significant differences between the plant-specific PRA models and the NRC's SPAR models, the staff undertook a major effort to further enhance the SPAR models; this effort went beyond the original anticipations of the pilot program. The results of these RES activities are the subject of this report.

Finally, the third major objective was to perform inspections in accordance with the NRC's Temporary Instruction (TI) 2515/149 (Ref. 4). This objective was undertaken by the sites' Resident Inspectors and Senior Resident Inspectors, as well as the Senior Reactor Analysts from the NRC's regional offices. This activity included an item-by-item verification of many of the tasks performed by the licensees, although not on all systems on all plants. (Refer to Ref. 5 for a full description of inspection activities and findings.)

As previously noted, 20 nuclear power plant units participated in the pilot program. The following list identifies the 20 units by region and plant:

Region I	Region II	Region III	Region IV
Hope Creek	Surry 1 & 2	Braidwood 1 & 2	Palo Verde 1, 2 & 3
Limerick 1 & 2		Prairie Island 1 & 2	San Onofre 2 & 3
Millstone 2 & 3			South Texas 1 & 2
Salem 1 & 2			

These units represent a reasonable cross-section of U.S. plant types, ages, designs, and reactor manufacturers; however, the pilot did not include any Babcock & Wilcox (B&W) reactors. For complete vendor coverage, it would have been useful to include B&W plants; however, a more important consideration was the availability of internal events, Level-1 at-power plant PRAs, and the varied experience of the licensees' staffs to exercise the models. The NEI has stated that, in this regard, the pilot participants were a reasonable representation of industry capabilities.

3.2   Activities and Schedule

During the first half of 2002, the NRC held a series of public meetings and workshops, in order to set the stage for the initiation of the pilot in the second half of 2002. The 3-day workshop in Chicago in July 2002 provided an opportunity to identify system boundaries and active components, attain familiarity with data reporting requirements, and resolve site-specific issues. Then, on August 28, 2002, the NRC published Regulatory Issue Summary 2002-14 to notify addressees that the agency was beginning a 6-month pilot program on September 1, 2002. The NRC then issued program guidelines (Ref. 1), which were attached to Regulatory Issue Summary 2002-14, Supplement 1, dated September 30, 2002. In addition, to guide the inspections of licensee submittals, the NRC issued Temporary Instruction (TI) 2515/149, "Mitigating Systems Performance Index Pilot Verification" (Ref. 4), on September 24, 2002.

Licensees began collecting data in September 2002, and they began sending their monthly submittals of the NEI electronic worksheets to the NRC via email in October 2002. Most licensees submitted their final worksheets in March 2003 (for the previous month); however, several participants voluntarily elected to continue their submittals for a short time thereafter. The NRC's inspections of licensee MSPI submittals (in accordance with TI 2515/149) took place throughout the late fall of 2002 and into the first quarter of 2003. However, because of the incomplete state of some licensees' MSPI-related documentation, not all pilot plants received full inspections.

In January 2003, the NRC held a public workshop to provide a mid-course assessment of the pilot, identify technical and process issues that had surfaced over the 3 preceding months, and adjust the pilot program accordingly. A number of the technical issues would require resolution prior to full implementation of the MSPI. Consequently, over a period of 6 months, the RES staff worked to fully assess the implications of the issues and to provide recommendations for their resolution. In May 2003, the staff issued a White Paper on the MSPI methodology for review and comment (Ref. 6). That White Paper provided background material for the staff to use in briefing the NRC's Advisory Committee on Reactor Safeguards (ACRS), Subcommittee on Reliability and PRA and Subcommittee on Plant Operations, on July 8, 2003.

The ACRS Subcommittees generally had no major concerns with either the MSPI concept or the direction of the research.

On July 23, 2003, RES presented the details and technical basis for modifications to the MSPI methodology. Industry representatives then held a workshop on August 20, 2003, to exercise the proposed changes, and the RES and NRR staffs observed the exercise.

In August 2003, the RES staff completed its enhancement of the 11 unique SPAR models (for all 20 nuclear units in the pilot). The results of this task are discussed in Appendix B.

The staff completed the pilot in September 2003, but some analyses beyond the original scope continued for several more months. For example, Appendix K to this report presents the results of sensitivity studies regarding the effect of data input errors and valve omissions on MSPI results The staff continues to address other issues, many of which relate to MSPI implementation. (Discussion and resolution of implementation issues are beyond the scope of this report.)

4. RESULTS OF INDEPENDENT VERIFICATION

The purpose of the independent verification of the MSPI was to obtain reasonable assurance of the adequacy of the inputs into the calculation and the reasonableness of pilot plant results. The RES staff accomplished this objective by assessing the individual inputs to the MSPI calculation on a plant-by-plant, system-by-system, and (in many cases) component-by-component basis. In addition, the staff compared the MSPI results using the plant-specific PRA models and the NRC's SPAR resolution models. Thus, this project included detailed independent verification of the following factors:

- baseline data
- current performance data
- FV/UA and FV/UR importance measures
- electronic spreadsheet calculations
- overall MSPI results

Toward that end, the RES staff reviewed the pilot plant data submittals to identify train-specific unavailability baselines, and they tabulated the results by system and train type. The verification generally showed that the pilot plant submittals were reasonable. However, the verification did identify several baseline UAs that were lower than the unplanned UA values listed in Table 1 of Appendix F to draft NEI 99-02 (Ref. 2). Clearly, the sum of two positive values cannot be less than any individual value. The effort also identified situations where the average planned UA based on a 3-year period could result in a baseline that is too high. This could arise if an unusually long planned train outage occurred in this baseline period. Additional guidance may be needed in this area.

The staff also reassessed the baseline failure rates from Table 2 of Appendix F to draft NEI 99-02. These "generic" industry-wide failure rates are common to all pilot plant submittals and, hence, this task was not *per se* a verification of pilot plant data submittals. Consequently, the reassessment of component failure rates led to a substantially expanded related task. The basic conclusion is that the current values in Table 2 of Appendix F to NEI 99-02 are not truly representative of component performance in 1995 – 1997 as supposed, and are not appropriate for use in the MSPI. Section 5.1 and Appendix C to this report provide additional discussion on this matter.

As part of the independent verification, the RES staff tabulated and compared current UA results for the 3-year period both across plants and with baselines. This comparison did not identify any current UA entries as outliers. The staff also compared the pilot plant unreliability data with data searches of the Equipment Performance and Information Exchange (EPIX) using the NRC-developed Reliability and Availability Database System (RADS) software. In general, this comparison revealed that the results across the 20 pilot plants for the number of failures and demands (or hours) were comparable to the results obtained from EPIX and RADS. However, the staff did identify inconsistencies between EPIX data and MSPI data on a plant-specific basis. In addition, the verification effort identified pilot plant data entry errors, including cases of double- or "multiple-counting" of failures or demands. The staff brought these errors to the attention of the licensees, and most were corrected by the final submittals in March 2003.

The RES staff also verified the pilot plant FV/UA and FV/UR values by comparing them with results obtained from the existing SPAR Rev. 3 models. Those SPAR models had previously been benchmarked against licensee models and, in most cases, were within a factor of 2 to 3 of licensee PRAs for core damage frequency. However, with regard to risk model importances at the component level, the staff found significant discrepancies between existing SPAR Rev. 3

models and the corresponding plant PRA models. Therefore, the staff initiated an additional SPAR enhancement to help resolve these differences for the 11 SPAR models that cover the 20 pilot plants. The limited SPAR-versus-plant PRA resolution effort succeeded in identifying and resolving many issues related to component FVs. Using the geometric mean (over all monitored components at a plant) as the figure-of-merit, the SPAR resolution models agreed with the 11 unique plant PRA FV/URs within a factor of 2 on average. Internal events CDF results also were also comparable. However, significant differences were discovered for certain components, especially those in the SWS and CCW where initiating event contributions to importance may be relevant. Appendix B to this report discusses the results of this effort in detail.

With regard to FV/UA and FV/UR values submitted by the pilot participants, the staff identified many instances in which the participants did not provide any FV/UR values for components that would otherwise be monitored. In general, these were low risk-significant components, but the participants did not provide the basis for omission. In addition, the staff identified some inconsistencies in UA for standby versus normally running trains. Additional guidance may be warranted in both of these areas.

As part of the verification, the staff verified the MSPI calculations within the NEI spreadsheet by comparing the results with those obtained using an independently developed spreadsheet. Results from both spreadsheets agreed.

The final step in the independent verification involved comparing MSPI results from the pilot plant submittals and SPAR resolution models. Toward that end, the staff compared the results obtained using similar (previously verified) performance data but different risk parameters (CDF, FV/UA, FV/UR) from SPAR and plant PRAs. In the process, the staff had to correct some half-dozen data entry errors from the pilot plant spreadsheets. Nonetheless, overall, the MSPI results from the pilot plant submittals and the SPAR resolution models were found to be in very good agreement:

- In terms of color indications, the pilot plant models and SPAR resolution models yielded comparable (if not identical) results for the 4th Qtr 2002, depending on whether the frontstop is used or the effect of common-cause failure modeling is accounted for.

- For MSPI values above the practical limit of significance (1×10^{-7}), the numerical results obtained using the SPAR resolution models generally agree with those from the pilot plant risk models to within a factor of 3.

- The pilot plant results indicated positive MSPIs for 39 out of 100 systems, compared to 37 using the SPAR resolution risk model.

Appendix A to this report presents additional details concerning this verification activity.

Because of concerns regarding the adequacy of plant PRAs for use in the MSPI, and because of significant differences in importance measures derived from the licensees' PRA models and the SPAR models, the staff performed a detailed analysis that went beyond the original scope of the pilot. Specifically, the analysis investigated the effect of PRA model differences on the MSPI results, as described in Appendix B to this report. The procedure involved identifying major modeling differences between the licensees' PRA models and SPAR models, grouping the differences, creating "change sets" for the PRA computer code, requantifying the entire risk model, deriving new importance measures, and assessing how the newly derived importances could affect the MSPI indications and numerical results.

The detailed analysis of the sensitivity of MSPI results to differences in PRA modeling demonstrated that these differences should be manageable. For all 11 unique PRA models, only 3 issues could have a potentially significant impact on MSPI results. The study found that significant differences in major model inputs (such as system success criteria or initiating event frequencies) are the source of significant quantitative differences, whereas differences in basic event probabilities on the order of a factor of 2 to 3 have a much lesser effect on MSPI results. However, prior to MSPI implementation, it may be necessary to identify and resolve potentially significant modeling differences. (Developing such a process is beyond the scope of this report.)

Finally, while recognizing the fundamental differences between the MSPI, SDP, and SSU approaches, the RES staff compared these three measures to determine whether their results exhibited overall congruence. In this regard, the staff analyzed all 77 failures over 3 years, as reported in the MSPI program for all pilot plants. Toward that end, the staff evaluated the quarterly MSPI indication results that were measurably impacted by comparison to the equivalent SSU performance indications, as appropriate. When an SDP finding was available for the failure in question, the staff also compared these results. Appendix to this report presents additional detail concerning this analysis.

On the basis of its independent verification, the RES staff concluded that the MSPI appears to consistently provide a better measure of integrated system performance than the SSU indicators, while minimizing both *false positive* and *false negative* likelihoods to the extent possible.

5. PROPOSED MODIFICATIONS TO MSPI METHODOLOGY

As a result of the pilot, a number of significant issues have arisen regarding the fundamental MSPI methodology described in the 2002 draft of NEI 99-02 (Ref. 2). Resolution of these issues requires a thorough understanding of how each issue currently affects MSPI results plant-by-plant, within the group of pilot plants, and across the industry as a whole. Secondly, any proposed changes to the methodology must be carefully assessed to avoid introducing unintended consequences. Consequently, in assessing proposed changes to the MSPI methodology, the RES staff has made every attempt to compare results before and after the change whenever possible. In many cases, the staff was able to derive a direct quantification of the change (e.g., number of failures to the GREEN/WHITE threshold). In addition, the staff often used two other techniques (a) to compare 4th quarter 2002 MSPI results for the pilot plants with and without the proposed change, and/or (b) to use numerical simulations in which certain input parameters (such as the number of component failures in a plant system within a 3-year period) were assumed to have a degree of randomness. (See Appendix L to this report.) Sections 5.1– 5.6 describe six major issues and proposed modifications to the MSPI methodology.

5.1 Baseline Performance Data

Table 2 of Appendix F to NEI 99-02 provides the generic industry failure rates as currently used in the MSPI. These failure rates, and the derived "a" and "b" values, are used for baseline component unreliabilities, and as priors for the Bayesian update of current performance. An important principle behind the selection of generic baseline data for the MSPI is that industry performance in 1995 – 1997 has been deemed acceptable by NRC policy.

Contrary to this understanding, a closer review of the sources of reliability data in Table 2 revealed that the data actually reflect performance from the 1970s, 1980s, and early 1990s. Notably, failure rates for these time periods have been shown to be greater than current rates by factors of 2 to 4. This is significant because higher failure rates would skew the baseline unreliability terms in the MSPI formulation to much higher values. This, is turn, could bias MSPI results in a more negative and nonconservative direction. Unfortunately, the sources of data for 1995 – 1997 are incomplete. Thus, a major data collection and analysis effort would be required to derive accurate failure rates for that 3-year period.

By contrast, as discussed in Appendix C to this report, good sources of data are available for the 3-year period from 1999 through 2001. Comparison of the derived failure rates from the various sources indicates that it is possible to determine failure rates with reasonable accuracy. Furthermore, statistical trend analyses of EPIX data and LERs used in updated system reliability studies generally indicate no significant trend from 1995 through 2001. Even if one were to assume a trend despite those analyses, failure rates for 1999 – 2001 would be perhaps 20 percent less than those for 1995 – 1997, with wide scattering from one component failure rate type to the next. Finally, the staff calculated 4th quarter 2002 MSPI results for the pilot plants using both the 1999 – 2001 data, and rates extrapolated to 1996. While numerical MSPI results for individual systems varied from one case to the next, the staff detected virtually no difference in GREEN and WHITE indications between these two data sets, across all MSPI systems and all pilot plants. The overall conclusion is that reliability data for 1999 – 2001 are reasonably representative of performance in 1995 – 1997 and, therefore, can be used with virtually no difference in results.

Recommendation #1: *Table 2 of Appendix F to NEI 99-02 should be revised to use industry failure rates derived for 1999 – 2001 (given in Table C.2 of this report) as a surrogate for 1995 – 1997.*

5.2   Use of Frontstop to Address Invalid Indicators

Some system indicators associated with the MSPI have significant "false positive" issues. That is, for statistical reasons, there is a significant probability that a plant system at baseline performance will cross over the GREEN/WHITE threshold. Within the MSPI pilot program, these indicators have been called "overly sensitive indicators" or, in the extreme case, "invalid indicators." "Invalid indicators" are MSPIs that have the property that just one failure above baseline during the observation period can cause the index to go "WHITE." Appendix F to NEI 99-02 (Ref. 2) states this property as follows:

> If, for any failure mode for any component in a system, the risk increase (ΔCDF) associated with the change in unreliability resulting from single failure is larger than 1.0×10^{-6}, then the performance index will be considered invalid for that system.

As discussed in detail in Appendix D to this report, random failures that occur at a rate consistent with the industry performance are not indicative of a performance issue. That is, one failure over a 3-year performance monitoring period, or one failure above the normal expectation, can be argued not to constitute a significant trend. Thus, expected performance variation should not result in the crossing of a performance threshold.

Based on pilot plant submittals, the RES staff estimates that some 17 to 24 percent of systems have at least one component failure mode (e.g., standby motor-driven pump fail-to-start) that could be considered an invalid indicator. There also appears to be a strong correlation between high importance measure (Birnbaum, in particular) and the likelihood of being an invalid indicator. Thus, the need to balance a high rate of "true positives" (correctly identifying degraded performance) while minimizing "false positives" is the driver behind the "frontstop." Originally conceived as a firm limit on the minimum number of failures necessary for a component type to indicate WHITE, the concept evolved to become a "risk cap" on the *single* most risk-significant failure within a system in the 3-year reporting period. The "risk cap" meets all of the desired characteristics:

- Address invalid indicators (thereby reducing false positives).

- Be compatible with, but not ignore, the Unavailability index contribution.

- Maintain sensitivity (without adversely impacting false negatives).

Furthermore, the risk cap approach is consistent with Regulatory Guide 1.177, "An Approach for Plant-Specific, Risk-Informed Decisionmaking: Technical Specifications," dated August 1998 (Ref. 3), with regard to what constitutes an acceptable guideline for a small risk increase attributable to a permanent Technical Specification change.

The risk cap is the minimum of 5×10^{-7} and the change in the unreliability index (delta URI) associated with the *single* most risk-significant failure within a system in the 3-year reporting period. It applies only to the GREEN/WHITE threshold. By assigning a "risk cap" of 5×10^{-7}, the outcome has the following attributes:

- No single failure alone results in a WHITE indication.

- Two significant failures (each with a risk contribution greater than 5×10^{-7}) would very likely result in a WHITE indication.

- One significant failure with other less-significant failures could exceed the GREEN/WHITE threshold.

- One significant failure with a significant UAI contribution could exceed the GREEN/WHITE threshold.

- A situation in which the URI is near zero but the UAI is greater than 1×10^{-6} would result in a WHITE indication.

No other potential solutions to the invalid indicator issue that were considered had all of these desired attributes. Moreover, many options added even greater complexity than the risk cap.

In practice, with the use of the risk cap, the determination of which components constitute "invalid indicators" would be a moot point. The risk cap would always be applied to the delta URI associated with the *single* most risk-significant failure, as long as the delta URI is less than 1×10^{-5}. Because of the concern with failures that could potentially result in a delta URI that is greater than 1×10^{-5} (i.e., YELLOW), and the much greater risk significance attached to YELLOW over WHITE, the risk cap would not be applied to the WHITE/YELLOW or YELLOW/RED thresholds. Pilot plant results did not identify any such situations where a single failure resulted in a delta URI greater than 1×10^{-5}. In fact, the staff identified only a few components among the pilot plants for which *two* failures could result in a delta URI greater than 1×10^{-5}. However, a single failure that results in a delta URI greater than 1×10^{-5} cannot be ruled out for the rest of the industry.

If implemented as proposed in this section and Appendix D to this report, the recommended "frontstop" concept could obviate the need to have invalid indicators defer exclusively to the inspection process as originally proposed.

Recommendation #2: *A "frontstop" (as described in Appendix D to this report) should be used to address the "invalid indicator" issue. The frontstop would take the form of a risk cap of 5×10^{-7} on the change in the unreliability index (delta URI) associated with the **single** most risk-significant failure, so long as the delta URI is less than 1×10^{-5}. The frontstop would only be applied to the GREEN/WHITE threshold.*

5.3   Use of Backstop to Address Insensitive Indicators

Although the systems selected for monitoring are relatively risk-significant at most plants, the Birnbaum importances (Bs) for specific system *trains* at some plants may be relatively small numbers. This is attributable, in part, to the system selection process; that is, an indicator defined for systems that are important at many plants, but not all plants, may be *insensitive* at some plants. A low value of train B can also easily arise in highly redundant systems, and failure of *individual* trains in a highly redundant system may not yield a high conditional CDF, even if failure of the entire system would do so. In such cases, a large number of failures would be needed to produce a change in the MSPI that is greater than 1×10^{-6}/yr.

Those components for which a large number of failures would be needed to produce a change in the MSPI that is greater than 1×10^{-6}/yr have come to be called "insensitive indicators." However, what constitutes a "large" number of failures can be subjective. For the sole purpose of performing sensitivity studies to identify possible solutions to the insensitive indicator issue, the staff used a condition of "more than 20 failures" in the original definition. By this measure, approximately 11 percent of the systems for the 20 pilot plants had *all components within the system classified as insensitive indicators*. In addition, not unexpectedly, the staff found that the number of failures to WHITE for a component type inversely correlated with FV/UR. Thus, the residual heat removal (RHR) system was most likely to have an insensitive indicator, owing to its generally low risk-importance at power.

The occurrence of an unexpectedly large number of failures implies a performance issue that could well be cross-cutting (i.e., could affect other systems) and have a net effect on ΔCDF that is somehow not captured in the current MSPI calculations. Therefore, it is desirable to supplement the 1×10^{-6} threshold criterion for entry into "WHITE" with another criterion. This supplemental criterion will be based on the statistical significance of the observed number of failures, relative to prior expectations. When the observed number of failures is greater than or equal to a specified "backstop" value, a WHITE will be declared, independent of the calculated change in the MSPI. As discussed in Appendix E to this report, the staff has formulated the "backstop" threshold to have the following properties:

- The false positive rate will be low. This criterion can be formulated to say that the conditional probability of declaring "WHITE," given normal performance, will be very low.

- Of all the positives that occur under baseline conditions, only very few are *false* positives.

In essence, the backstop should rarely be invoked (in comparison to the calculated MSPI using the algorithm). Numerical simulations have confirmed that the backstop has this property.

Conceptually, the "backstop" is a limit on the total number of failures, of all failure modes and all components of one type in one system of a single nuclear power plant unit. Each system and type of component corresponds to a single backstop, with all failure modes combined. If the number of failures seen in the 3-year performance period is greater than or equal to the backstop number, the system or component has reached or exceeded the backstop and is denoted as "WHITE."

Using the method discussed in Appendix E to this report, the staff derived two types of backstops. The first is a generic set of backstops by component type. If the number of failures of similar components within a system (e.g., both emergency diesel generators or EDGs) **reached or exceeded** the backstop in a 3-year period, the system would be declared WHITE, regardless of the calculated MSPI. This would ensure that the system would not remain GREEN despite a large number of failures. The second approach allows a variable backstop based on the "expected number of failures" of similar components in a system over a 3-year period. The advantage of the variable backstop approach is that it allows for variation in design configuration (number of components), testing frequency, and operation. Given the large variations across the industry, the variable backstop approach is strongly encouraged.

With the "backstop" so defined, it is now possible to re-define what constitutes an "insensitive" system. If all component failure modes within a system require more failures to WHITE using the 1×10^{-6} criterion than the corresponding variable backstop counts, that system is defined as "insensitive." By this measure, approximately 33 percent of the 100 systems in the 20 pilot plants would be deemed "insensitive" (not including the adjustment for common-cause failure).

Recommendation #3: The variable "backstop" (as described in Appendix E to this report) should be used to address the "insensitive indicator" issue.

5.4   Treatment of Common-Cause Contribution to Fussell-Vesely

The original draft guidance for MSPIs (Ref. 2) included the following statement concerning common-cause failure:

> Some aspects of mitigating system performance cannot be adequately reflected or are specifically excluded from the performance indicators in this cornerstone. These aspects include... the effect of common-cause failure....

This approach would relegate regulatory oversight of potential common-cause failure (CCF) entirely to the inspection processes. Given a CCF-induced multiple failure, this would be analyzed under the SDP. However, this approach is not intended to proactively address the existence of conditions that promote CCF. The RES staff believes it is desirable to reflect the CDF significance of *all* performance changes that can validly be reflected in the MSPI, given the purpose of the MSPI and the character of the performance data and available models.

Most CCF models represent the CCF contribution to risk as being essentially proportional to overall failure probability. In such models, if the measured UR increases and the proportionality constants are left alone, the assessed CCF contribution increases along with the independent failure contribution. The RES staff approached MSPI quantification with the notion that a change in UR increases the change in core damage frequency (delta CDF) both through the independent failure contribution and through a CCF contribution. (The original draft approach would not add the CCF contribution.) Therefore, for a given data set and a given model, the approach proposed by the RES staff estimates a larger CDF change than the original draft approach in NEI 99-02 (Ref. 2). In many cases, this leads to a substantially lower number of failures to reach the GREEN/WHITE threshold.

Because the purpose of the MSPI is to flag *potential* performance problems based on operating experience, it seems most reasonable to propagate changes in observed UR through the parametric CCF model, and include the change in CCF contribution in the assessed change in CDF. If there is an underlying performance issue causing a real increase in UR, it may well relate to CCF anyhow.

Appendix F to this report provides a methodology for adjusting the MSPI Unreliability index terms proposed by NEI to address the CCF contribution to these indices. Specifically, Appendix F addresses the impact of a change in the independent failure probability on the CCF probability. The approach to address the CCF contribution provides a first-order mathematical approximation. It utilizes only one input beyond those already required by the MSPI, namely, the Fussell-Vesely (*FV*) importance value of the CCF event associated with each in-scope common-cause group. The increase in the URI term will vary depending on the common-cause importance of the component in question, the degree of coupling between total and common-cause failure rates, and the degree of redundancy of the component type.

Sensitivity studies described in Appendix F indicate that the net effect of CCF on the increase in URIs could range from as low as 5 percent for low degrees of redundancy (e.g., two-fold) and coupling, to an order-of-magnitude increase where the degree of redundancy is high (e.g., four-fold) and the coupling is strong. In a separate calculation for one particular failure mode for a highly redundant system at one pilot plant, the estimated number of failures-to-WHITE over a 3-year period went from more than 30 failures with no adjustment for CCF, to about 5 or 6 with CCF effects included.

The RES staff performed additional calculations to estimate the impact of CCF on the issue of invalid indicators, as well as the number of WHITE indicators that might result from the long-term implementation of the program. The percentage of pilot plant systems having at least one component failure mode with invalid indication increased from 17 percent without CCF to 24 percent with the effect of CCF included. However, with the use of the "frontstop," the matter of invalid indicators would be moot. It would simply mean that the effect of CCF would be to apply the "frontstop" more often than if CCF were not considered.

Likewise, the staff performed a numerical simulation of the likely outcome of including CCF in the revised MSPI formulation. (See Appendix L to this report.) That simulation indicated including the CCF effects might yield about one-third more WHITE indicators (compared to the results without CCF). The RES staff believes that this potential effect on the projected number of WHITE indicators is reasonable, given all of the limitations and approximations in the MSPI formulation. Moreover, the inclusion of CCF would substantially reduce the insensitive indicator issue, and minimize the need to rely on the performance-based "backstop." Using the revised definition of an insensitive system (i.e., one in which the number of failures to WHITE exceeds the "backstops" for all components in the system), the percentage of insensitive systems for the pilot plants drops from 33 percent to 20 percent when common-cause is considered.

Exercises performed by a number of pilot plant participants at the NEI workshop on August 20, 2003, indicated that detailed guidance and training would be required to implement the proposed inclusion of Fussell-Vesely importances for CCF. The exercises also revealed that, in some instances, common-cause modeling includes a complicated coupling of pumps, motors, breakers, and other components. Thus, participants found it difficult to determine the CCF-related FV importances. As a result, the RES staff has provided an alternative approach to address CCF in Appendix F to this report. This alternative approach allows the use of generic multipliers on the FV from independent failures as an appropriate adjustment to account for the effect of CCF.

Recommendation #4: *The MSPI formulation should include the common-cause failure contribution to FV importance (as described in Appendix F to this report), and NEI 99-02 should provide substantial guidance on the process for including this contribution.*

5.5   Exclusion of Active Valves Based on Birnbaum Importance

Appendix F to NEI 99-02 provides clarifying notes concerning the criteria for identifying those components that are to be monitored in the MSPI. NEI 99-02 provides specific guidance for valves, whether in series or parallel, for multi-train systems. That guidance is prescriptive in nature and is intended to ensure, to a first order of approximation, that important valves within a system are included. The expectation is that the number of valves to be monitored should not differ significantly from the number of pumps in all the systems monitored (i.e., about 20 valves) However, the pilot revealed that, in some cases, licensees would have to monitor as many as 46 valves. This far exceeds expectations and can pose a significant data collection burden, with no clear benefit in return.

Based on an analysis of all of valves monitored by the 20 pilot plants, it is possible to exclude low-importance valves without affecting the overall results of the MSPI. The Birnbaum importance measure has been deemed appropriate given that it is the measure directly used in calculating URI, and URI is the figure-of-merit of interest here. The analysis described in Appendix G to this report shows a cutoff B value of 1×10^{-6}/yr to significantly reduce burden, while still yielding reasonably conservative results. *The common-cause contribution to FV (and Birnbaum) must be added to the valve Birnbaums before the cutoff is applied.*

An important consideration is whether some minimum number of valves should remain in scope, regardless of their risk importance. Monitoring too few valves with the MSPI could have undesirable consequences. Firstly, the URI is more sensitive to failures of valves within a smaller population, and more likely to result in a false WHITE for a small number of failures. Secondly, valves that are not monitored with the MSPI could be subject to the inspection process. Thirdly, as the plant-specific PRA model changes to reflect changes in plant design or equipment performance, it is likely that importance measures will also change. Therefore, it seems reasonable to ensure that a minimum number of valves within fluid systems are monitored, regardless of their risk significance.

Recommendation #5: *The guidance in Appendix F to NEI 99-02 should be revised to allow licensees the option to exclude low risk valves with Birnbaum importance measures (adjusted for common-cause effects) less than 1×10^{-6}/yr (as described in Appendix G to this report).*

The guidance in Appendix F to NEI 99-02 should also be revised to add appropriate cautions regarding the potential negative consequences of monitoring too few valves within a system. Also, the decision to use this option should be made at the beginning of the system boundary identification, and not changed unless a major PRA model revision causes significant movement of valve Birnbaums above or below the cutoff.

5.6   Contribution of Support System Initiators to Fussell-Vesely Importance

Of the six systems within the scope of the MSPI, the service water system (SWS) and component cooling water (CCW) are the two that could serve in the dual roles of both supporting other systems when called upon, and initiating a transient if the SWS or CCW is lost entirely or substantially degraded.

All PRA models provide risk measures such as Fussell-Vesely importance, risk achievement worth (RAW), and Birnbaum importance from basic event probabilities for SWS and CCW components. However, while all of the models include the components' contributions from the "support system" role of the SWS and CCW, some models do not include the contributions from the loss of the SWS or CCW as an initiating event. This is because the initiating event frequencies used in some plant-specific PRAs have been based on plant and/or industry experience, and use explicit values for the frequencies. A given frequency may use a distribution with mean and variance, but the calculated value may, in some way, be separate from the linked PRA model. In other models, the PRA analyst may choose to link a loss of SWS initiator fault tree directly into the PRA computer model. Either approach is acceptable, so long as it is based on valid equipment performance data, accounts for the potential for common-mode failure based on plant-specific characteristics and design, and is generally consistent with industry operating experience.

All other things being equal, a plant PRA model that uses initiator fault trees explicitly for loss of SWS and/or CCW (where importance of the initiating event components is accounted for) will result in higher Fussell-Vesely (FV) and Birnbaum risk measures for an associated basic event than a model that uses a point-estimate frequency. The difference between the two approaches would be a function of the importance of the given initiator to the overall calculated CDF, as well as the importance of the particular component (and basic event) within the SWS or CCW of interest. During the MSPI workshop on January 21, 2003, the staff surveyed the pilot plant participants and determined that plant PRA models fell into three categories, including (a) those that used fault trees for loss of SWS and CCW initiators that were directly linked in the PRA model, (b) those that used fault trees and/or event trees outside of the linked PRA model to quantify the frequencies, which were manually entered into the PRA model in the same manner as a medium loss-of-coolant accident (LOCA) frequency, and (c) those that used frequencies based on industry experience, updated with plant-specific data. Category "a" is the most prevalent, with about two-thirds of the pilot plants using this approach. These differences in approach clearly create an inconsistency for the MSPI methodology, which relies heavily on using calculated risk measures (FV divided by basic event probability), rather than (for example) a requantification of the entire PRA model.

Some pilot plant analysts have performed sensitivity studies to determine the importance of including the contribution of support system initiators in the FV risk measure. Toward that end, the analysts first performed calculations using the existing linked fault tree initiator models, and then repeated the calculations with the fault tree initiator essentially turned off. Differences in FV using the two approaches can be expected to be strong functions of the following factors:

- importance of the initiator to overall CDF
- importance of the component within the system
- system configuration and design
- importance of recovery actions and success criteria

At the lower end, the differences in calculated FV with and without initiator fault trees were less than 1 percent. At the upper end, differences as high as an order of magnitude in FV were seen for some components. The contribution of SWS and CCW components to FV both as initiators and mitigators need to be included if the full risk importance is to be properly accounted for.

Clearly, if the safety-related CCW and/or SWS to be monitored in the MSPI are strictly standby systems, their loss cannot initiate a plant transient. The already-calculated FV values for the components of these systems are proper and no further action is necessary.

Assuming that no initiator fault trees exist, it is possible to avoid the need to include the initiators' contributions to FV if all CCW/SWS components to be monitored in the MSPI have a Birnbaum (maximum for all failure modes) importance of less than 1×10^{-6}/yr. Conversely, it is only necessary to account for the initiators' contributions to FV if none of the above conditions are met.

In the proposed resolution, licensees would have two options. Those plant PRA models that do not use fault trees for loss of SWS and/or CCW could either (a) add such fault trees and recalculate the FV importance measures, or (b) use an approximation to adjust the FV to account for the contribution in a manner that is proportional to the importance of the system initiator to CDF, and proportional to the importance of the component within the system, as described in Appendix H. This adjustment is shown to be conservative, yielding from zero to approximately 25 percent higher FV (based on regression analysis) than would be expected using an initiator fault tree. Given this potential conservatism in the approximation to adjust the FV, licensees may well choose to develop initiator fault trees for loss of SWS and CCW for the purpose of the MSPI.

Recommendation #6: *The guidance in Appendix F to NEI 99-02 should be revised to require the inclusion of the contribution of cooling water support system initiators to FV importance (as described in Appendix H to this report).*

As discussed in Appendix H, one option to address this issue would be to add initiator fault trees for loss of the SWS and/or CCW. A second option would be to use an approximation to conservatively adjust the FV to account for the contributions from support system initiators.

5.7   Additional Issues for Resolution

Finally, it should be noted that the NRC staff has not yet resolved all issues identified during the course of the pilot program; however, recommendations 1–6 (above) address the major *technical* issues associated with the proposed MSPI formulation. The staff continues to address other issues, which largely relate to the *implementation* of the MSPI. In addition, the guidance in Appendix F to the draft NEI 99-02 continues to be modified to incorporate findings resulting from this research. Finally, it should be noted that a separate task group is developing a process to identify and resolve potentially significant modeling differences between the licensees' PRA models and the NRC's SPAR models.

6.   REFERENCES

1. H.G. Hamzehee, et al., U.S. Nuclear Regulatory Commission (NRC). NUREG-1753, "Risk-Based Performance Indicators: Results of Phase 1 Development." NRC: Washington, DC. April 2002.

2. Nuclear Energy Institute (NEI). NEI 99-02 (Draft Report), "Regulatory Assessment Performance Indicator Guideline," Section 2.2 ("Mitigating Systems Performance Index") and Appendix F ("Methodologies for Computing the Unavailability Index, the Unreliability Index, and Determining Performance Index Validity"). NEI: Washington, DC. 2002.

3. U.S. Nuclear Regulatory Commission (NRC). Regulatory Guide 1.177, "An Approach for Plant-Specific, Risk Informed Decisionmaking: Technical Specifications." NRC: Washington, DC. August 1998.

4. U.S. NRC. Inspection Manual, Temporary Instruction (TI) 2515/149, "Mitigating Systems Performance Index Pilot Verification." NRC: Washington, DC. September 24, 2002.

5. U.S. NRC. Temporary Instruction (TI) 2525/149, "Mitigating Systems Performance Index Pilot Verification," ADAMS Accession #ML030840341. NRC: Washington, DC. March 25, 2003.

6. U.S. NRC. Interoffice Memorandum from Scott F. Newberry (RES/DRAA) to John A. Zwolinski (NRR), "Request for Review of Mitigating Systems Performance Indices White Paper," Adams Accession #ML03135020B and ML031360121. NRC: Washington, DC. May 12, 2003.

APPENDIX A. SUMMARY OF MSPI VERIFICATION EFFORT

Appendix A
Summary of MSPI Verification Effort

A.1   Introduction

The Mitigating Systems Performance Index (MSPI) is a measure of approximate change in core damage frequency (CDF) resulting from changes in mitigating system component unreliability performance and train unavailability. The MSPI was evaluated for six mitigating systems at each pilot plant, with cooling water support systems combined into a single indicator. For each mitigating system, the MSPI equation is as follows:

$$MSPI = \left(CDF_P\right)\left(\sum \frac{FV_F}{UR_F}\right)\left(UR_C - UR_B\right) + \left(CDF_P\right)\left(\sum \frac{FV_P}{UA_P}\right)\left(UA_C - UA_B\right) \quad \text{(Eq. A.1)}$$

where $MSPI$ = ΔCDF for the system (from changes in component UR and train UA)
CDF_P = internal events core damage frequency per calendar year (from plant PRA)
FV_P = Fussell-Vesely importance measure of the component or train (from plant PRA)
UR_P = component unreliability (from plant PRA)
UR_C = current component unreliability (Bayesian update using data from most recent 3 years)
UR_B = baseline component unreliability (Table 2 from Appendix F of draft NEI 99-02)
UA_P = train unavailability (from plant PRA)
UA_C = current train unavailability (data from most recent 3 years)
UA_B = baseline train unavailability (1999 – 2001 plant experience for planned and industry average for unplanned; Table 1 from Appendix F of draft NEI 99-02).

The first summation in Equation A.1 is over all monitored components within the system, while the second summation is over all trains within the system.

The MSPI calculation requires various inputs, including monitored components within each monitored system, train UA_Bs and component UR_Bs, train and component performance during the rolling 3-year data collection period (train UA_C, component failures, and associated demands or run hours), and risk model importance information (CDF_P, FV_P/UR_P, and FV_P/UA_P). The MSPI verification discussed in this appendix addressed most of these inputs. However, inspection efforts covered the determination of monitored components and the collection of plant performance data, so those areas are not addressed here. Also, the risk model importance information is discussed in Appendix B, which addresses the development of standardized plant analysis risk (SPAR) resolution models.

MSPI verification results presented in this appendix are based on the pilot plant data submittals dated March 21, 2003. However, several plants submitted corrected data in early April 2003. Also, modifications were made to Surry 1 and 2 in September 2003 to remove the internal flooding contribution to the model, consistent with the other pilot plants and the intent of the MSPI. Those corrections were included in the verification effort.

A.2   MSPI Baseline Data Verification

For train baseline unavailabilities, draft NEI 99-02 indicates that plant-specific and train-specific planned outages over a 3-year period should be used to develop train-specific planned UA baselines. Unplanned UA baselines are industry-average values (over the period 1999 – 2001) listed in Table 1 from Appendix F of draft NEI 99-02. The plant- and train-specific UA baselines listed in the pilot plant data submittals therefore include both planned and unplanned UA. The actual UA data by quarter reported by the MSPI pilot plants includes the combined sum of plant-specific planned and plant-specific unplanned UA.

As part of the verification effort, the pilot plant data submittals were reviewed to identify train-specific UA baselines. Results were tabulated by system and train type. An example of the results of this effort is presented in Table A.1, covering the emergency alternating current (AC) power system. Several observations were noted based on this tabulation of UA baselines (which was performed for all five types of systems):

(1) Several train baseline UAs were lower than the unplanned UA values listed in Table A.1 from Appendix F of draft NEI 99-02. According to the NEI guidelines, no train UAs should be lower than the values listed in that table.

(2) Additional guidance is needed for cases where the baseline period for establishing UA planned includes an unusually long train outage (as might have occurred for emergency diesel generator B at Hope Creek, in Table A.1). In such cases, the resulting baseline may be too high, and results from the other trains may be more appropriate in terms of expected baseline performance.

(3) The use of different UA baselines for similar trains within a system, especially if only a 3-year period is used to establish the baseline, may imply differences between trains that do not actually exist.

(4) Industry average results for UA planned baselines (using 1999 – 2001 data) may be more appropriate than plant-specific, train-specific results obtained over a 3-year period.

For component baseline failure rates, values from Table 2 in Appendix F of draft NEI 99-02 are to be used. Appendix C addresses the applicability of those Table 2 baseline failure rates to the MSPI pilot program. The results of that comprehensive review are that a set of new failure rates (Year 2000) should be used based on industry performance during the period from 1999 through 2001. The MSPI verification results in this appendix are based on use of the Year 2000 failure rates.

A.3   MSPI Current Performance Data Verification

Current UA results for the period July 1, 1999 through June 30, 2002 (3Q99 – 2Q02) for the 20 pilot plants were tabulated as shown in Table A.1. Results were compared across plants and with baselines to identify any suspect values. No current UA entries were identified as outliers.

To verify the pilot plant unreliability data, 3-year results (3Q99 – 2Q02) were compared with data searches of the Equipment Performance and Information Exchange (EPIX) using the NRC-developed Reliability and Availability Database System (RADS) software. An example of the type of comparison made is presented in Table A.2 for the emergency AC power system, emergency diesel generator failure to start. In general, the results at the overall level (summation of all 20 pilot plants) for numbers of failures and demands (or hours) are comparable to the results obtained from EPIX/RADS. However, for individual plants, the failure counts may not agree (e.g., the pilot plant data indicate a failure, while EPIX/RADS do not, or conversely). Also, the demand (or hour) totals may be significantly different. These plant-specific inconsistencies between EPIX data and MSPI data should be resolved at some point, especially if EPIX or the consolidated data entry (CDE) program is to be used in the future to submit data for the MSPI.

The comparison between pilot plant unreliability data and EPIX/RADS results identified several potential pilot plant data entry errors. An example of such errors was "multiple counting" of component demands, where component demands were summed over several components and then the sum was reported as the result for each individual component. (The NEI pilot plant data sheet would then again sum these values to obtain an overall demand total for the component type, resulting in multiple, incorrect counting of component demands.) Another example involved reporting of emergency diesel generator failures occurring during the load or run phase. One plant appeared to report failures during this phase as both failure to load and run (FTLR) and failure to run (FTR). Also, one plant appeared to report a single failure as occurring every quarter during the 3-year period, thereby over counting the failures by a factor of 12. Most of the potential data entry errors were corrected in the March 21, 2003 pilot plant data submittals.

A.4   MSPI FV/UA and FV/UR Verification

Pilot plant FV/UA and FV/UR values were verified by comparing with results obtained from SPAR models. The existing SPAR Rev. 3 models had been benchmarked against licensee models and were, in most cases, within a factor of 2 to 3 of licensee PRAs for core damage frequency. However, with regard to risk model importances at the component level, significant discrepancies were found between existing SPAR Rev. 3 models and the corresponding plant PRA models. (There were often large differences between the SPAR Rev. 3 estimates for FV/UA and FV/UR and those from the pilot plant risk models.) Therefore, an additional SPAR enhancement effort was performed to help resolve these differences for the eleven SPAR models that cover the 20 pilot plants. The results of that effort are discussed in Appendix B, covering the development of SPAR resolution models. Using the SPAR resolution models, FV/UA and FV/UR comparisons with pilot plant risk model results usually agreed within a factor of 3 for the more risk-significant components (FV/UR or FV/UA > 1.0, or Birnbaum > 1.0×10^{-5}/year). However, several of the SPAR resolution models contain success criteria, basic event values, or initiating event frequencies (chosen to match the plant risk models) that would not be allowed under current SPAR development guidelines. These issues will need to be addressed before the SPAR resolution models can be issued as official SPAR models.

Several miscellaneous issues were identified with regard to FV/UR and FV/UA values. One is that a significant number of pilot plants did not list such values for some of their monitored components. It was not clear whether these components were not included in the risk model or these components were lost in the risk model truncation process. Guidelines might need to be developed to cover such instances. Another type of issue involves modeling of multiple-train systems with one or more trains normally running. In such cases, the risk models often assume certain trains are normally running and the others are standby. Then, train UA (from planned and unplanned outages) is included only for the standby train(s). Risk model FV/UA values obtained from such a model need to be modified to accurately reflect operations where any of the trains can be normally running (or standby). Such modifications were identified for two- and three-train systems, but additional guidance may be needed for other types of configurations. Note that these modifications to FV/UA values were made to the risk models during the MSPI verification process.

A.5   MSPI Spreadsheet Calculation Verification

As part of the MSPI verification effort, the MSPI calculations performed within the NEI spreadsheet (used by the pilot plants to report their data) were verified by comparing results from an independently developed spreadsheet. Results from both types of spreadsheets agreed.

A.6   MSPI Results Verification

The final step in verifying pilot plant MSPI results was a comparison of ΔCDF results with those obtained using SPAR resolution model results. For these comparisons, both approaches used the same pilot plant performance data (with several corrections listed below), the same baseline UA (pilot plant values) and UR (Year 2000 values recommended in Appendix C), and the same mission times (24 hours for all systems except the emergency diesel generators, and eight hours for the emergency diesel generators). However, the pilot plant MSPI results used the pilot plant risk model values for CDF, FV/UA, and FV/UR (with changes made by the plant during the SPAR enhancement efforts), while the SPAR MSPI results used SPAR resolution model values.

Several potential data corrections were included in this MSPI comparison:
(1) Surry 2 emergency alternating current (EAC) power system [failure to run (FTR): 4 failures reduced to 0]
(2) Salem 1 SWS [motor-driven pump (MDP) failure to start (FTS): 17 failures reduced to 0]
(3) Millstone 3 high-pressure safety injection (HPSI) (MDP FTR: 8080 hours reduced to 80.8)
(4) Limerick 2 residual heat removal (RHR) system (missing data filled in with Limerick 1 data)
(5) Prairie Island 1 and 2 CCW (changed standby MDP to running MDP).

Because of the changes listed above (UR baselines, data corrections, mission times, FV/UA, and FV/UR), the pilot plant MSPI values listed in this appendix are different from those calculated in the March 21, 2003 pilot plant submittals. The changes were made to more accurately reflect current assumptions and methodologies.

Presented in Table A.3 are the MSPI results (ΔCDF and performance color) for the 3-year data period ending December 31, 2002 (4Q02), using the pilot plant risk models. Three MSPIs out of 100 are greater than 1.0×10^{-6}/yr and are therefore WHITE. However, with the proposed frontstop, the Palo Verde 2 heat removal system (HRS) and Salem 1 EAC MSPIs drop below 1.0×10^{-6}/yr and are GREEN. This leaves only one WHITE for the quarter, Braidwood 1 HRS. This MSPI is WHITE because of two diesel-driven pump failures to start and one failure to run over the 3-year period.

Presented in Table A.4 are the same MSPIs for the same period, but calculated using the SPAR resolution model values for FV/UR, FV/UA, and CDF. Two of the 100 MSPIs are WHITE using the SPAR resolution models. However, with the proposed frontstop, the Salem 1 EAC MSPI drops below 1.0×10^{-6}/yr and is GREEN. This leaves only one WHITE, Braidwood 1 HRS. Therefore, with the proposed frontstop, both the plant PRA and the SPAR resolution models indicate one WHITE, Braidwood 1 HRS. (Note that this result represents a snapshot of the MSPI for only one quarter during the pilot, and is not inclusive of all other possible WHITE indications during other quarters).

For MSPI values above 1.0×10^{-7}/year (the practical limit of significance), the SPAR resolution model results generally agree with the plant risk model results to within a factor of 3. This is expected, because the SPAR resolution model development generally resulted in FV/UR and FV/UA values that were within a factor of 3 of the plant risk model results.

Overall, the plant risk model MSPI results (Table A.3) include 40 positive ΔCDF entries and 60 negative values. Also, the average MSPI value is 1.0×10^{-8}/year, which is essentially neutral. In comparison, the SPAR resolution risk model MSPI results (Table A.4) include 37 positive and 63 negative entries. The average MSPI value in the SPAR resolution model is -2.7×10^{-8}/year, also neutral.

A.7 Summary of MSPI Verification Effort

The MSPI verification effort involved the comparison of plant risk model parameters (FV/UR, FV/UA, and CDF) with corresponding SPAR risk model values. The verification effort also included comparison of MSPI ΔCDF results obtained from the two risk models. In general, the existing SPAR Rev. 3 models did not match the plant risk models with respect to the FV/UR, FV/UA, and CDF parameters. An additional SPAR enhancement effort was required in order to develop SPAR resolution models that produced FV/UR and FV/UA values within a factor of 3 of the plant risk model values. Given these SPAR resolution models, the MSPIs calculated are in general agreement with the plant risk model MSPI results. Specifically, for most MSPIs with a ΔCDF greater than 1.0×10^{-7}/year, the SPAR resolution model results generally agree with the plant risk model results within a factor of 3.

Table A.1 MSPI Pilot Plant Emergency Diesel Generator UA Baselines and Current Performance Summary

Pilot Plant Data (3Q99 - 2Q02)
8/24/2003

Emergency AC (EAC) System

Pilot Plant	# Trains	UA Current Performance (3Q99 - 2Q02)				UA Train Baseline (1999 - 2001)				Site Current Average	Site Baseline Average	Comments
		DGA	DGB	DGC	DGD	DGA	DGB	DGC	DGD			
Braidwood 1	2.0000	0.0112	0.0124			0.0122	0.0122			0.0086	0.0122	
Braidwood 2	2.0000	0.0039	0.0069			0.0122	0.0122					
Hope Creek	4.0000	0.0093	0.0122	0.0110	0.0148	0.0107	0.0958	0.0132	0.0155	0.0118	0.0338	DGB baseline is much too high
Limerick 1	4.0000	0.0197	0.0129	0.0106	0.0134	0.0241	0.0154	0.0069	0.0098	0.0100	0.0119	
Limerick 2	4.0000	0.0032	0.0066	0.0044	0.0095	0.0116	0.0048	0.0119	0.0109			
Millstone 2	2.0000	0.0129	0.0120			0.0156	0.0149			0.0125	0.0153	
Millstone 3	2.0000	0.0090	0.0104			0.0130	0.0138					
Palo Verde 1	2.0000	0.0067	0.0087			0.0039	0.0050			0.0076	0.0049	
Palo Verde 2	2.0000	0.0124	0.0052			0.0083	0.0023					
Palo Verde 3	2.0000	0.0070	0.0057			0.0039	0.0059					
Prairie Island 1	2.0000	0.0099	0.0092			0.0195	0.0189			0.0136	0.0148	
Prairie Island 2	2.0000	0.0123	0.0231			0.0084	0.0126					
Salem 1	3.0000	0.0081	0.0126	0.0089		0.0090	0.0109	0.0086		0.0091	0.0093	
Salem 2	3.0000	0.0091	0.0083	0.0073		0.0091	0.0101	0.0082				
San Onofre 2	2.0000	0.0241	0.0193			0.0254	0.0234			0.0199	0.0189	
San Onofre 3	2.0000	0.0165	0.0194			0.0124	0.0144					
South Texas 1	3.0000	0.0178	0.0155	0.0172		0.0161	0.0160	0.0143		0.0171	0.0160	
South Texas 2	3.0000	0.0136	0.0138	0.0245		0.0164	0.0168	0.0166				
Surry 1	2.0000	0.0234	0.0250			0.0224	0.0167			0.0270	0.0224	
Surry 2	2.0000	0.0333	0.0261			0.0310	0.0194					
Average	Current	0.0130										
	Baseline	0.0149										
		0.0132	without Hope Creek DGB									

Table A.2 MSPI Pilot Plant Unreliability Data Comparison with EPIX/RADS for Emergency Diesel Generator Failure to Start

Comparison of Pilot Plant Data (3Q99 - 2Q02) with EPIX/RADS Data
6/10/2003
Pilot plant data as of 3/21/03. EPIX database including 4Q02, as accessed using RADS.

System	Component	Failure Mode	Pilot Plant				EPIX/RADS			Comments
			Pilot Plant	# Components	# Failures	# Demands	# Hours	# Failures	# Demands	# Hours	
EAC	EDG	FTS	Braidwood 1	2	1	116		1	87		
			Braidwood 2	2	0	123		0	112		
			Hope Creek	4	0	192		0	200		
			Limerick 1	4	0	227		0	198		EPIX estimate for test demands (12) is inaccurate. EPIX FTLR demands used.
			Limerick 2	4	0	201		0	198		EPIX estimate for test demands (12) is inaccurate. EPIX FTLR demands used.
			Millstone 2	2	0	106		0	92		EPIX estimate for test demands (41) is inaccurate. EPIX FTLR demands used.
			Millstone 3	2	0	77		0	92		EPIX estimate for test demands (23) is inaccurate. EPIX FTLR demands used.
			Palo Verde 1	2	1	72		0	149		
			Palo Verde 2	2	0	72		0	152		
			Palo Verde 3	2	1	72		0	152		
			Prairie Island 1	2	0	74		0	74		
			Prairie Island 2	2	2	92		1	95		
			Salem 1	3	0	216		0	212		
			Salem 2	3	0	216		0	246		
			San Onofre 2	2	0	72		0	72		
			San Onofre 3	2	0	72		0	72		
			South Texas 1	3	0	108		0	147		EPIX estimate for test demands (30) is inaccurate. EPIX FTLR demands used.
			South Texas 2	3	0	108		0	153		EPIX estimate for test demands (31) is inaccurate. EPIX FTLR demands used.
			Surry 1	1.5	1	159		2	98		Both plant data and EPIX include data from swing EDG.
			Surry 2	1.5	1	158		3	49		Plant data include data from swing EDG. EPIX does not include the swing EDG.
			Totals	49	7	2533		7	2650		
			Failure Rate (Jeffreys noninformative prior)			2.96E-03			2.83E-03		

Table A.3 Pilot Plant MSPI Results for the 4th Quarter 2002

Plant MSPI Results 4th Quarter 2002

Year 2000 Baselines, 8-hr EDG Mission Time

Licensees' Plant PRA Model	Mitigating System				
	EAC	HPI	HRS	RHR	SWS/CCW
Braidwood 1	-9.58E-08	4.39E-08	2.28E-06	1.51E-08	6.13E-08
Braidwood 2	-1.62E-07	-2.00E-08	1.22E-07	1.71E-07	6.99E-08
Hope Creek	2.95E-07	5.61E-07	4.88E-07	-1.73E-09	-6.66E-08
Limerick 1	-5.90E-08	-5.90E-08	-6.68E-08	-3.95E-08	-1.87E-08
Limerick 2	-2.13E-07	-1.13E-07	-1.11E-07	-8.10E-08	2.24E-08
Millstone 2	-4.59E-07	-2.65E-07	-3.91E-07	3.75E-10	6.37E-07
Millstone 3	-4.67E-07	-2.63E-07	-8.78E-07	-8.18E-08	1.04E-07
Palo Verde 1	1.10E-07	2.42E-08	-5.37E-07	-8.30E-09	-8.00E-08
Palo Verde 2	-5.23E-08	1.35E-08	3.02E-06	-6.01E-09	-1.02E-07
Palo Verde 3	1.79E-07	2.38E-08	-3.59E-07	-4.01E-09	-1.49E-07
Prairie Island 1	-2.03E-07	-8.48E-09	-1.14E-07	-7.65E-08	3.76E-07
Prairie Island 2	3.62E-07	-1.03E-08	-1.90E-08	2.59E-08	2.89E-07
Salem 1	2.84E-06	-8.34E-09	-4.03E-07	-3.30E-07	1.39E-07
Salem 2	-3.17E-06	4.20E-08	-2.51E-07	-9.79E-08	6.33E-07
San Onofre 2	-2.29E-08	-1.47E-08	-8.42E-07	-2.42E-08	-9.06E-08
San Onofre 3	2.87E-09	-4.42E-07	-9.52E-07	-2.44E-08	-4.29E-07
South Texas 1	1.01E-07	-5.72E-08	-6.97E-07	4.58E-08	1.07E-08
South Texas 2	6.05E-08	2.02E-07	2.74E-07	5.22E-08	-1.67E-07
Surry 1	3.91E-07	-5.94E-09	-3.17E-08	-7.93E-09	1.97E-07
Surry 2	3.03E-07	-3.41E-09	-3.44E-08	-2.12E-10	1.92E-07

Note: With the proposed frontstop, the Palo Verde 2 HRS and Salem 1 EAC MSPIs become GREEN. However, Braidwood 1 HRS remains WHITE. Also, note that these results are a snapshot in time, representing the MSPI for 4Q2002 only.

Table A.4 SPAR Resolution MSPI Results for 4th Qtr 02

SPAR Resolution MSPI Results 4th Quarter 2002

Year 2000 Baselines, 8-hr EDG Mission Time

SPAR Resolution Model	Mitigating System				
	EAC	HPI	HRS	RHR	SWS/CCW
Braidwood 1	-1.57E-07	8.50E-08	2.58E-06	3.95E-11	4.09E-08
Braidwood 2	-2.49E-07	-1.86E-08	3.39E-07	1.41E-07	7.88E-08
Hope Creek	4.54E-07	7.54E-07	6.39E-07	2.46E-08	-1.81E-08
Limerick 1	-1.80E-07	-1.02E-07	-1.44E-07	-1.14E-07	-9.44E-09
Limerick 2	-2.20E-07	-1.01E-07	-1.24E-07	-1.22E-07	8.96E-09
Millstone 2	-1.63E-06	-2.59E-07	-1.07E-06	4.42E-07	4.16E-07
Millstone 3	-9.52E-08	-1.56E-07	-7.34E-07	-8.08E-08	-3.42E-08
Palo Verde 1	6.44E-08	8.09E-09	-5.37E-07	-3.08E-09	-1.36E-07
Palo Verde 2	-1.54E-07	3.38E-09	7.27E-07	-2.51E-09	-1.56E-07
Palo Verde 3	2.48E-07	1.13E-08	-3.54E 07	-2.65E-09	-1.75E-07
Prairie Island 1	-1.40E-07	5.82E-09	-8.93E-08	-5.68E-08	2.11E-07
Prairie Island 2	2.40E-07	-2.33E-09	-4.70E-08	9.82E-09	1.73E-07
Salem 1	4.13E-06	-9.34E-09	-1.01E-06	-2.07E-07	4.00E-07
Salem 2	-4.60E-06	4.19E-08	-3.89E-07	-8.20E-08	5.20E-07
San Onofre 2	-1.03E-07	-1.81E-09	-9.01E-07	-2.48E-08	-1.72E-08
San Onofre 3	-2.43E-08	-4.78E-07	-9.56E-07	-2.26E-08	-2.45E-07
South Texas 1	-2.50E-07	-7.72E-09	-4.41E-07	4.44E-09	1.67E-08
South Texas 2	-2.59E-07	1.34E-08	-8.95E-08	5.08E-09	-3.68E-08
Surry 1	6.59E-07	-1.76E-08	-3.01E-08	-3.86E-09	6.34E-07
Surry 2	3.67E-07	-9.28E-09	-2.59E-08	-4.47E-10	4.93E-07

Note: With the proposed frontstop, the Salem 1 EAC MSPI becomes GREEN. However, Braidwood 1 HRS remains WHITE. Also, note that these results are a snapshot in time, representing the MSPI results for 4Q2002 only.

APPENDIX B. SUMMARY OF SPAR ENHANCEMENT EFFORT

Appendix B
Summary of SPAR Enhancement Effort

B.1   Introduction

As part of the Mitigating Systems Performance Index (MSPI) pilot program, the standardized plant analysis risk (SPAR) models developed by the U.S. Nuclear Regulatory Commission (NRC) were used to verify the adequacy of plant probabilistic risk assessment (PRA) inputs to the MSPI. The MSPI is a measure of approximate change in core damage frequency (CDF) resulting from changes in mitigating system component unreliability performance and train unavailability. The MSPI is evaluated individually for five indicators consisting of six mitigating systems at each pilot plant. For each mitigating system, the MSPI equation is the following:

$$MSPI = \left(CDF_P\right)\left(\sum \frac{FV_r}{UR_r}\right)\left(UR_C - UR_B\right) + \left(CDF_P\right)\left(\sum \frac{FV_P}{UA_P}\right)\left(UA_C - UA_B\right) \quad \text{(Eq. B.1)}$$

where *MSPI* = *ΔCDF for the system (from changes in component UR and train UA,*
  CDF$_P$ = *internal events core damage frequency per calendar year (from plant PRA)*
  FV$_P$ = *Fussell-Vesely importance measure of the component or train (from plant PRA)*
  UR$_P$ = *component unreliability (from plant PRA)*
  UR$_C$ = *current component unreliability (Bayesian update using data from most recent 3 years)*
  UR$_B$ = *baseline component unreliability (Table 2 from Appendix F of draft NEI 99-02)*
  UA$_P$ = *train unavailability (from plant PRA)*
  UA$_C$ = *current train unavailability (data from most recent 3 years)*
  UA$_B$ = *baseline train unavailability (1999 – 2001 plant experience for planned and industry average for unplanned; Table 1 from Appendix F of draft NEI 99-02).*

The first summation in Equation B.1 is over all monitored components within the system, while the second summation is over all trains within the system.

To verify the adequacy of plant PRA inputs to the MSPI, the plant PRA CDF, FV/URs, and FV/UAs were compared with corresponding values from the SPAR models. [The terms UR$_C$, UR$_B$, UA$_C$, and UA$_B$ in Equation B.1 are independent of the plant PRA and were therefore covered under separate verification efforts.] Several types of SPAR model comparisons were made: SPAR Rev. 3 model as obtained from the SAPHIRE Users' Group website, SPAR resolution model, and SPAR resolution model but with selected basic event and initiating event values associated with various modeling issues (termed SPAR issues) changed back to the SPAR recommended values. Each plant's SPAR Rev. 3 importance measures that were used to generate MSPI inputs were compared with plant PRA results. Where significant differences were noted, modeling changes were identified that would resolve some of these differences. Modifications that were deemed within the SPAR development guidelines and additional modifications not within the SPAR development guidelines were added to obtain the SPAR resolution model. Examples of modifications not within the guidelines include basic event probabilities significantly different from the SPAR development guideline values, human error probabilities not derivable from the SPAR human error methodology, initiating event frequencies significantly different from SPAR development guidelines, and system success criteria that may not be appropriate. The impacts of these modeling issues on the SPAR resolution model results were also evaluated.

The SPAR Rev. 3 models represent an upgrade from the older SPAR Rev. 3i models. These upgrades were generated mainly using information obtained during plant visits as part of the Significance Determination Process (SDP) verification effort. (Additional basic event data upgrades were also part of this process.) However, when the resulting SPAR Rev. 3 models were first compared with plant PRA FV/UR values submitted as part of the MSPI pilot plant effort, significant differences were noted for many of the plants. (Wherever FV/UR is discussed, FV/UA is also included.) This was somewhat surprising, but previous SPAR upgrade efforts typically were not focused on importance measures for individual components. Therefore, additional effort was expended to identify and resolve the differences, and this effort led to the development of the SPAR resolution models.

Finally, many of the MSPI pilot plant PRA models include initiating event fault trees for loss of service water system (SWS) or component cooling water (CCW) system. When evaluating the importances of components within those two systems, many of those plant PRAs include importance contributions from the component to both the initiating event and the mitigating system. SPAR models do not presently include initiating event fault trees. Therefore, FV/UR comparisons for components within the SWS and CCW may be misleading. (In general, the SPAR FV/UR values should be lower than the plant values, because the SPAR results do not include importance resulting from the initiating event.)

B.2 SPAR Resolution Model Development and Comparison Process

The MSPI pilot program includes 20 commercial nuclear power plant units. However, because of similar units at some sites, eleven individual SPAR models cover these 20 units. (For example, the SPAR model for Braidwood covers each of the two units at that site.) For each of these eleven SPAR Rev. 3 models, a comparison spreadsheet was developed covering all of the monitored components for the plant in question. Table B.1 presents the spreadsheet developed for Braidwood Units 1 and 2. The plant PRA FV/URs were then listed for each monitored component. As a starting point, the SPAR Rev. 3 FV/URs were compared with the plant PRA values. This comparison is presented in Table B.1 as ratios of SPAR Rev. 3 value divided by plant PRA value. For these ratios, a value of 1.0 indicates agreement between the SPAR value and the plant PRA value.

To develop the final SPAR resolution model, plant PRA cut sets were compared with SPAR cut sets. Reasons for differences were identified and appropriate changes were then made to the SPAR Rev. 3 model and/or the plant PRA. (Note that this detailed comparison of cut sets at times led to changes in the plant PRA.) After several changes were made to the SPAR Rev. 3 model, a new comparison of FV/URs was made using the spreadsheet (under the SPAR resolution column). Although this process could be extended almost indefinitely, the SPAR resolution model development was truncated when most, if not all of the FV/UR ratios (for components with FV/URs > 0.1) lay between 0.3 and 3.0. The final SPAR resolution model results were then loaded into the spreadsheet and individual and summary comparison results generated. In Table B.1 the SPAR resolution FV/UR ratios are significantly closer to 1.0 than the SPAR Rev. 3 model results. For example, the auxiliary feedwater (AFW) diesel-driven pump (DDP) (SPAR event AFW-DDP-FR-1B) FV/UR ratio drops from 8.05 (8.05 times higher FV/UR than the plant PRA value) to 0.97. Also shown in the table are the geometric averages of ratios for components with FV/URs > 1.0 and within the range 1.0 to 0.1. Again, the geometric averages for the SPAR resolution model are much closer to 1.0 compared with the

SPAR Rev. 3 results. Finally, the standard deviations of the FV/UR ratios within these two ranges are also shown.

As noted in Section B.1, the SPAR resolution models include some basic event values not typically allowed under the SPAR development guidelines. Therefore, the effects of these values on the SPAR resolution results were evaluated. Because many basic events can be involved, a standardized set of issue categories was developed. (These issue categories are listed in Table B.2 and discussed in more detail in Section B.4.) Basic event data changes were then grouped within these issue categories. As an example, Table B.3 shows the basic events grouped within each of the applicable issue categories for Braidwood. Note that the power-operated relief valve (PORV) issue in Table B.3 is not a basic event data change, but a model structure change involving the success criterion for the power-operated relief valves. The impact of each issue category on the SPAR resolution results was then determined by changing all of the basic events within the issue category back to SPAR recommended values and rerunning the SPAR model. Resulting Birnbaum ratios were then compared with the SPAR resolution results to determine how much of an impact that issue category had. Table B.2 presents the results of this type of sensitivity analysis for Braidwood. Note that Table B.2 uses Birnbaum importances, which are more informative because the Birnbaum importance incorporates not only the FV/UR portion of Equation B.1 but also the CDF factor. (The Birnbaum is just the CDF times the FV/UR.) When using Birnbaums, the component Birnbaum ranges of interest are $> 1.0 \times 10^{-5}$/year and 1.0×10^{-5} to 1.0×10^{-6}/year. A review of the SPAR issue results in Table B.2 indicates that the PORV success criterion (one-of-two for the SPAR resolution and plant PRA models, and two-of-two for the SPAR Rev. 3 model) most affects the Birnbaum results, especially for components with Birnbaums $> 1.0 \times 10^{-5}$/year. However, modeling of direct current (DC) power also has a significant impact.

To evaluate the potential impacts of these Birnbaum importances (from the various model runs) on actual MSPI ΔCDF results, two additional types of comparisons were performed. The first used actual pilot plant data for the period 2000 – 2002 (termed the 4Q2002 data set) to evaluate the UR_C and UA_C terms in Equation B.1. The system MSPIs were then calculated using Birnbaums obtained from each model run. Note that these MSPI calculations used the Year 2000 recommended baseline unreliability values discussed in Appendix C, and did not include the effects of common-cause modeling. Results from this type of comparison for Braidwood 1 and 2 are presented in Table B.4. (Results are presented for each unit in Table B.4 because the plant data — component failures and demands or hours and train unavailabilities — are different for each unit.) Note that the plant PRA and SPAR resolution MSPI colors agree — all GREEN except for the Unit 1 auxiliary feedwater system (HRS in the table) WHITE. In contrast, the SPAR Rev. 3 model predictions result in a Unit 1 HRS YELLOW and a Unit 2 HRS WHITE. Finally, changing the SPAR resolution PORV success criterion from one-of-two (the plant PRA criterion) to two-of-two (the SPAR recommended criterion) is the only issue category that results in color changes compared with the SPAR resolution (and plant PRA) results.

The MSPI comparisons using the 4Q2002 data set are highly dependent upon the actual system failures that occurred during that interval. For example, if the Braidwood 1 HRS had not experienced several diesel-driven pump failures, then the sensitivity of the MSPI to the PORV success criterion would not have been identified in the analysis presented in Table B.4. Therefore, a second type of MSPI ΔCDF comparison was also performed. This comparison postulates an additional component failure above the expected number of failures in the 3-year period, with other component types and the train unavailabilities within the system postulated to be performing at their baseline conditions. This evaluation is performed separately for each component type and failure mode within a system. Results for Braidwood 1 are summarized in Table B.5. (Results for Braidwood 2 would be slightly different, because of differences in component demands and hours and train unavailabilities.)

Rather than a ratio (used in Tables B.1 and B.2), a difference factor is used as the measure of agreement in Table B.5. The difference factor is defined as the following:

$$\text{Difference factor} = (\Delta MSPI_{SPAR} - \Delta MSPI_{Plant\ PRA})/1.0 \times 10^{-6}/year \quad \text{(Eq. B.2)}$$

The logic behind the difference factor is the desire to express SPAR model sensitivities in terms of absolute impacts on MSPI ΔCDF predictions. A ratio, as used in Tables B.1 and B.2, could be misleading. For example, if the plant PRA MSPI prediction were 1.0×10^{-8}/year, a ratio of 3 (SPAR MSPI prediction divided by plant PRA prediction) would indicate that the SPAR MSPI prediction is 3.0×10^{-8}/year, or higher than the plant result by 2.0×10^{-8}/year. However, if the plant PRA MSPI prediction were 1.0×10^{-6}/year, then a ratio of 3 indicates the SPAR MSPI prediction is 3.0×10^{-6}/year, or higher than the plant result by 2.0×10^{-6}/year. This second example is clearly much more important in terms of impacts on the MSPI, even though both examples have a ratio of 3. Finally, the denominator of 1.0×10^{-6}/year in Equation B.2 is used to conveniently express results in terms of 1.0×10^{-6}/year units. For the two examples just discussed, the difference factors would be 0.02 and 2.0, respectively, clearly indicating the greater impact of the second example. For difference factor comparisons, a value of 0.0 indicates agreement between the SPAR and plant PRA results.

B.3   Summary of SPAR Resolution Model Results

Detailed results of the comparisons between the SPAR resolution model and the plant PRA results are presented in Tables B.1, B.2, B.4, and B.5 for Braidwood, as discussed in Section B.2. Similar tables were generated for the other ten SPAR models but are not presented in this appendix. However, summary statistics for each comparison are presented in Tables B.6 through B.8. Throughout the discussion of summary statistics in this section, it should be kept in mind that individual component results can vary significantly, even if the summary statistics indicate good overall agreement.

Table B.6 summarizes the CDF and Birnbaum comparisons. This table is a summary of the information presented in Table B.2 for Braidwood, but including all eleven SPAR models. The CDF ratios presented in the table are the SPAR model CDF divided by the plant PRA CDF. As indicated in the table, the SPAR Rev. 3 model CDF is an average of 1.63 times the corresponding plant PRA CDF. The worst agreement is for Braidwood, where the SPAR Rev. 3 CDF is 3.12 times the plant PRA CDF. However, the SPAR resolution models on average have a CDF 1.12 times higher than the corresponding plant PRA CDF. Also, the Braidwood ratio improves from 3.12 to 1.11.

In terms of component Birnbaum importances, Table B.6 presents summary statistics for two ranges of Birnbaums: > 1.0×10^{-5}/year, and 1.0×10^{-5} to 1.0×10^{-6}/year. The Birnbaum ratios presented are the SPAR model component Birnbaum divided by the plant PRA Birnbaum (average of all monitored components for the plant). For the more important components (Birnbaum > 1.0×10^{-5}/year), the SPAR Rev. 3 models on average predict importances that are 0.66 times the plant PRA values. Also, the average standard deviation is 2.26. In contrast, the SPAR resolution models predict component Birnbaums that are 1.27 times the plant PRA values, with an average standard deviation of 0.93. Therefore, the SPAR resolution models result in improved component Birnbaum predictions (compared with plant PRA values) in terms of both the average ratio and the average standard deviation.

For components with Birnbaums in the range 1.0×10^{-5} to 1.0×10^{-6}/year, the SPAR Rev. 3 models on average predict importances that are 1.08 times the plant PRA importances. The average standard deviation of these ratios is 3.53. In contrast, the SPAR resolution model average prediction is 1.36 times the plant PRA importance, with an average standard deviation of 1.98. For these less important components, the SPAR resolution models predict higher importances but the variability in predictions is reduced.

Table B.7 summarizes the MSPI comparison based on the 4Q2002 data set. This table is a summary of information presented in Table B.4 for Braidwood, but including the results from all 20 pilot plants. Shown in Table B.7 are the color comparisons (SPAR model versus plant PRA model) for the SPAR Rev. 3 and resolution models by system. Cases where the predictions do not agree are highlighted in the table. For the SPAR Rev. 3 models, 3 of the 100 cases do not agree in MSPI color. In all three cases, the SPAR result is more severe (e.g., WHITE rather than GREEN, or YELLOW rather than WHITE). For the SPAR resolution models, only 1 of 100 cases does not agree. Note that these comparisons do not include modifications to the MSPI predictions resulting from application of the proposed frontstop, backstop, or common-cause failure adjustments.

Finally, Table B.8 summarizes the MSPI comparisons based on the postulated additional failure above the baseline expected number of failures. This table is a summary of information presented in Table B.5 for Braidwood, but including all eleven SPAR models. Each difference factor entry in Table B.8 is an average of the results for the monitored components and failure modes for the plant in question. On average, the SPAR Rev. 3 models predict MSPIs (given one failure above the expected number of failures) that are 1.4×10^{-7}/year higher than the plant PRA predictions (a difference factor average of 0.14). However, the average of the standard deviations is 0.96, or 9.6×10^{-7}/year. This standard deviation is considered to be large. In comparison, the SPAR resolution models predict MSPIs that are an average 3.0×10^{-8}/year lower than the plant PRA predictions. However, the average standard deviation is much improved, from 0.96 (9.6×10^{-7}/year) to 0.26 (2.6×10^{-7}/year).

Difference factors summarized in Table B.8 provide some additional information concerning the SPAR resolution model development effort. In general, difference factors of 0.10 or smaller (impacts of 1.0×10^{-7}/year or smaller) indicate that differences between the SPAR model Birnbaums and the plant PRA Birnbaums do not significantly impact MSPI predictions. A review of summary information in Table B.8 indicates that on average the SPAR resolution effort for Limerick, Prairie Island, South Texas, and Surry had little impact on the MSPI predictions. For all of these plants, the SPAR Rev. 3 and SPAR resolution average difference factors and average standard deviations are small. However, a review of the Birnbaum comparison ratios in Table B.6 would not indicate that these SPAR resolution efforts had little impact. This reinforces the belief that the difference factor comparisons are the most meaningful in terms of evaluating SPAR models.

B.4   Summary of SPAR Model Issues

The following is a list of generic issues concerning the SPAR Rev. 3 models. This list was generated based on SPAR model development and comparison efforts before the SPAR resolution effort started. However, the SPAR resolution effort helped to reinforce the validity of the list.

Support System Initiating Event Fault Trees

Many plant PRAs model support system initiating events with fault trees that are then linked to the mitigating system fault trees when solving for sequence cut sets. This approach more correctly accounts for component importances (for those components in the affected systems) compared with the SPAR approach of using an initiating event frequency.

Initiating Event Frequencies

Differences in initiating event frequencies between the SPAR models and plant PRAs drive many of the differences observed in component importances. This is especially true for loss of SWS, CCW, and DC bus initiators, but is also true for other initiators. Present SPAR values are based mainly on industry average performance during the period 1987 – 1995. Industry performance has improved considerably since that period.

Reactor Coolant Pump (RCP) Seal Failure Modeling

The RCP seal failure modeling in SPAR, resulting from a loss of cooling differs from most plant PRAs. SPAR seal failure probabilities range from 0.7 to 0.08, while the plant PRAs often use 1.0 or a very low probability.

PORV Success Criterion during Feed and Bleed

Many plant PRAs require only one-of-two PORVs for success during feed and bleed. The SPAR models require two-of-two PORVs for success. This difference has a major impact on the Braidwood model results, and may significantly impact other plant models.

Loss of Offsite Power (LOOP) and Station Blackout (SBO) Modeling

Differences between plant PRAs and SPAR models with respect to LOOP and SBO include the following: preferential alignment of backup emergency power sources (assumed in order to simplify the models), modeling of dual unit LOOP, and offsite power recovery and emergency diesel generator mission time modeling. All of these can result in significant differences in component FVs.

Component Failure Rates

Significant differences can exist between plant PRA and SPAR component failure rates. The SPAR values are based mainly on published system study reports (1987 – 1993, 1995, or 1997, depending upon the study) and generic estimates (NUREG-1150, representing component performance before 1983). Again, significant performance improvement has occurred since the periods covered by these sources.

Steam Generator Tube Rupture (SGTR) Modeling

Significant differences between the plant PRA and SPAR models were noted with respect to the SGTR modeling. These differences are focused on the treatment of human actions in response to SGTR events including both the characterization of these actions and their values. For several plants these actions are dominant contributors and significantly impact the component FVs.

As noted in Section B.2, development of the SPAR resolution models included SPAR mode changes not typically allowed under current development guidelines. Many of the changes fall under one or more of the SPAR generic issues listed above. An example of such changes for the Braidwood model is presented in Table B.3. Similar tables were prepared for the other ten SPAR resolution models. In order to systematically and efficiently evaluate the sensitivity of SPAR model Birnbaum results to these basic event changes, a standard set of SPAR issue categories was developed. This set of SPAR issue categories is listed below:

(1) PORVs: power-operated relief valve success criterion

(2) ACP: alternating current (AC) power, including LOOP frequency, LOOP recovery and emergency diesel generators

(3) DCP: direct current (DC) power

(4) LOCAs: loss-of-coolant accidents, including reactor coolant pump seal leakage and stuck open relief valves

(5) HPI: high-pressure injection, including feed-and-bleed

(6) HRS: decay heat removal (auxiliary feedwater or reactor core isolation cooling)

(7) RHR: residual heat removal system

(8) SWS/CCW: service water or component cooling water systems, including initiating event frequencies

(9) PCS: power conversion system

(10) Misc.: other issues

These 10 SPAR issue categories are organized mainly by the system(s) affected. Other types of categories could have been chosen. For example, all human errors could have been grouped into a single category. Also, all initiating events could be included in a single category. The sensitivity effort described in this section covers only the system-related categorization scheme.

SPAR resolution model sensitivities to these issue categories were evaluated by replacing each basic event value (within a given issue category) with the SPAR Rev. 3 recommended value. New SPAR Birnbaums were then generated and their effects on MSPI predictions were determined. Summary results of this effort for all eleven SPAR resolution models are given in Tables B.9 and B.10, which present difference factor results assuming a single failure above the baseline expected number of failures. Table B.9 summarizes the average difference factor for each plant, while Table B.10 summarizes the standard deviation of the difference factor.

In Tables B.9 and B.10, the SPAR resolution model sensitivities to the SPAR issue categories can be classified based on three types of outcomes:

- Large impact: difference factor greater than 0.50 (5.0x10^{-7}/year), likely to result in an MSPI color change, given failures within a system
  Medium impact: difference factor between 0.10 and 0.50, with the potential to result in an MSPI color change given sufficient failures within a system

- Low impact: difference factor less than 0.10, unlikely to result in an MSPI color change.

In Tables B.9 and B.10, the large impact entries have been highlighted. Both tables indicate that the PORV success criterion issue has a large impact for Braidwood Units 1 and 2. The plant PRA assumes one-of-two PORVs is sufficient for feed and bleed, while the SPAR guideline requires two-of-two PORVs. However, this issue was not found to have a large or medium impact at any of the other applicable MSPI pilot plants.

Also, both tables indicate that loss-of-coolant accident (LOCA) issues have a large impact at Millstone 2. There are ten different basic (or initiating) event changes in the LOCA issue category for Millstone 2, covering initiating event frequencies, stuck open relief valve probabilities, and reactor coolant pump seal LOCA probability. Some values are higher for the SPAR Rev. 3 model and some are higher for the SPAR resolution model. Without reviewing each of the basic or initiating event changes individually, it is not clear which are driving the differences. Again, the LOCA issue category does not result in a large impact on MSPI predictions for the other pilot plants.

Finally, the SWS/CCW issue category has a large impact on MSPI predictions for Salem Units 1 and 2. In this case, the plant PRA has a loss of service water system initiating event frequency that is approximately 30 times lower than the SPAR Rev. 3 value. However, there are thirteen basic (or initiating) event changes in this issue category for Salem, so other events may also be contributing to the large impact. The SWS/CCW issue category does not result in a large impact on MSPI predictions for the other pilot plants.

Table B.11 summarizes the SPAR issue categories in terms of their impacts (large, medium, or small) on MSPI predictions. As discussed above, there are three cases where an issue category resulted in a large impact on MSPI predictions. Also, based on difference factor averages (Table B.9) or standard deviations (Table B.10), there are fifteen cases where an issue category resulted in a medium impact on MSPI prediction.

Table B.1 Comparison of SPAR Model FV/UR and FV/UA with Plant PRA Values (Braidwood 1)

Enhanced SPAR Model Development Results

Plant Unit	Critical Hours (3Q99 - 2Q02)	Date		Core Damage Frequency (note a)		
				Plant PRA	SPAR Rev. 3.02	SPAR Resolution
Braidwood 1	25394	11/13/2003	Per Critical Hour		1.11E-08	3.96E-09
			Per Calendar Year	3.01E-05	9.40E-05	3.35E-05
			Plant Critical Operation Availability	?	0.97	0.97
			SPAR CDF/Plant CDF		3.12	1.11

Information from Plant MSPI Data Submittal Spreadsheet

System	Component Type	Component Identifier	Component Description	SPAR Model SPAR Basic Event	Alternate Event	Plant PRA FV/UR or FV/UA (note b)	SPAR Rev. 3.02	SPAR Resolution/Plant (note e)
IIR3	MDP	1AF01PA	AF Pump 1A	AFW-MDP-FR-1A		16.60	0.33	0.36
HRS	Train (MDP)	AFA	Aux Feedwater Train A (TM)	AFW-MDP-TM-1A		14.90	0.25	0.37
HRS	DDP	1AF01PB	AF Pump 1B	AFW-DDP-FR-1B		4.16	4.37	0.98
HRS	Train (DDP)	AFB	Aux Feedwater Train B (TM)	AFW-DDP-TM-1B		2.91	8.05	0.97
RHR	MOV	1SI8811R	Charging Pump to Cold Leg Injection isol Valve	HPR-MOV-CC-8811B	HPR-MOV-CC-SMPB	1.83	0.10	0.73
RHR	MDP	1RH01PB	RH Pump 1B	RHR-MDP-FC-1B		1.82	0.14	0.85
RHR	Train (MDP)	RH1B	RH Pump 1B (TM)	RHR-MDP-TM-1B		1.76	0.08	0.67
CCW	MDP	1CC02PA	CC Pump 1A	CCW-MDP-FR-1A		1.57	0.07	0.23
RHR	MOV	1SI8811A	Charging Pump to Cold Leg Injection isol Valve	HPR-MOV-CC-8811A	HPR-MOV-CC-SMPA	1.11	0.16	0.48
RHR	MDP	1RH01PA	RH Pump 1A	RHR-MDP-FC-1A		1.04	0.24	0.70
RHR	Train (MDP)	RH1A	RH Pump 1A (TM)	RHR-MDP-TM-1A		1.02	0.12	0.37
HPI	MOV	1SI8804B	RH HX B to CV Pump suction isol valve	HPR-MOV-CC-RHRB		0.87	0.20	1.50
HPI	Train (MDP)	SIB	SI Pump Train 1B (TM)	HPI-MDP-TM-1B		0.84	0.00	0.01
EAC	EDG	DG1A	EDG 1A	EPS-DGN-FS-1A	EPS-DGN-FC-1A	0.83	0.61	1.32
HPI	MDP	1CV01PA	CV Pump 1A	CVC-MDP-FR-1A		0.67	0.10	0.80
HPI	MDP	1CV01PB	CV Pump 1B	CVC-MDP-FR-1B		0.67	0.01	0.20
EAC	Train (EDG)	DG1A	EDG 1A (TM)	EPS-DGN-TM-1A		0.66	0.52	0.98
SWS	MDP	1SX02PB	SX Pump 1B	ESW-MDP-FS-1B		0.63	0.56	1.27
SWS	Train (MDP)	SX1A	SX Pump 1A (TM)	ESW-MDP-TM-1A		0.21	0.48	1.35
SWS	Train (MDP)	SX1B	SX Pump 1B (TM)	ESW-MDP-TM-1B		0.21	0.48	1.35
EAC	EDG	DG1B	EDG 1B	EPS-DGN-FS-1B	EPS-DGN-FC-1B	0.40	0.60	1.90
CCW	MDP	1CC02PB	CC Pump 1B	CCW-MDP-FS-1B		0.26	0.06	0.41
SWS	MDP	1SX02PA	SX Pump 1A	ESW-MDP-FS-1A		0.17	1.26	5.02
EAC	Train (EDG)	DG1B	EDG 1B (TM)	EPS-DGN-TM-1B		0.14	0.76	2.34
HPI	MOV	1SI8801A	CV Pump to Cold Leg injection isol valve	CVC-MOV-CC-8801A		0.10	0.00	0.03
HPI	MOV	1SI8801B	CV Pump to Cold Leg injection isol valve	CVC-MOV-CC-8801B		0.10	0.02	0.68

B-9

Table B.1 Comparison of SPAR Model FV/UR and FV/UA with Plant PRA Values (Braidwood 1) (continued)

Enhanced SPAR Model Development Results

Plant Unit	Critical Hours (3Q99 - 2Q02)	Date		Core Damage Frequency (note a)		
				Plant PRA	SPAR Rev. 3.02	SPAR Resolution
Braidwood 1	25394	11/13/2003				
			Per Critical Hour		1.11E-08	3.96E-09
			Per Calendar Year	3.01E-05	9.40E-05	3.35E-05
			Plant Critical Operation Availability	?	0.97	0.97
			SPAR CDF/Plant CDF		3.12	1.11

Information from Plant MSPI Data Submittal Spreadsheet

System	Component Type	Component Identifier	Component Description	SPAR Model SPAR Basic Event	Alternate Event	FV/UR or FV/UA Ratio — Plant PRA (note b)	SPAR Rev. 3.02	SPAR Resolution/ Plant (note e)
HPI	MDP	1SI01PA	SI Pump 1A	HPI-MDP-FS-1A	HPI-MDP-FC-1A	0.09	0.00	0.00
HPI	MDP	1SI01PB	SI Pump 1B	HPI-MDP-FS-1B	HPI-MDP-FC-1B	0.09	0.03	0.08
HPI	MOV	1CV8804A	RH HX A to CV Pump suction isol valve	HPR-MOV-CC-RHRA		0.06	2.70	8.14
HPI	MDP	SIA	SI Pump Train 1A (TM)	HPI-MDP-TM-1A		0.06	0.00	0.00
RHR	MOV	1CC9412A	CC water from RH HX isol Valve	CCW-MOV-CC-RHRA		0.05	1.62	5.23
RHR	MOV	1CC9412B	CC water from RH HX isol Valve	CCW-MOV-CC-RHRB		0.05	1.84	21.09
CCW	MOV	1SX007	Unit 1 CC HX Outlet MOV	ESW-MOV-CC-1SX007		0.05	0.34	1.98
HPI	MOV	1CV112C	VCT Outlet isol Valve	CVC-MOV-OO-112C		0.05	0.31	2.81
HPI	MOV	1CV112E	RWST to CV Pump Suction Valve	CVC-MOV-CC-112E		0.05	0.36	4.30
HPI	MOV	1CV112B	VCT Outlet isol Valve	CVC-MOV-OO-112B		0.03	0.42	3.44
HPI	MOV	1CV112D	RWST to CV Pump Suction Valve	CVC-MOV-CC-112D		0.03	0.42	3.48
HPI	Train (MDP)	CVB	CV Pump Train 1B (TM)	CVC-MDP-TM-1B		0.03	0.13	2.27
CCW	MOV	0SX007	Unit 0 CC HX Outlet MOV	ESW-MOV-CC-0SX007		0.03	0.42	2.19
HPI	Train (MDP)	CVA	CV Pump Train 1A (TM)	CVC-MDP-TM-1A	CVC-MDP-TM-1B	0.02	0.21	3.62
RHR	MOV	1RH8716A	RH HX Discharge Crosstie Valve	Not modeled		0.01	0.00	0.00
RHR	MOV	1RH8716B	RH HX Discharge Crosstie Valve	Not modeled		0.01	0.00	0.00
CCW	Train (MDP)	CC1A	CC Pump 1A (TM)	CCW-MDP-TM-1A	CCW-MDP-TM-1B	0.00	9.83	67.46
CCW	Train (MDP)	CC1B	CC Pump 1B (TM)	CCW-MDP-TM-1B		0.00	9.83	67.46

Plant FV/UR-A >= 1.00		
Geometric Mean	0.29	0.56
Standard Deviation of Sample	2.58	0.26
Variance of Sample	6.65	0.07
1.00 > Plant FV/UR-A >= 0.10		
Geometric Mean	0.12	0.64
Standard Deviation of Sample	0.36	1.23
Variance of Sample	0.13	1.52

Note a - The plant PRA core damage frequency is for internal events without internal flooding

Note b - Entries highlighted in gray are changes to the MSPI data submitted for the plant (3/21/03 submittal) These are either changes made by the plant (PRA changes or other reasons) or changes judged by the MSPI/SPAR analysts to be appropriate

Note c - SPAR Rev 3 model on website as of 6/15/03

Note d - SPAR Rev 3 model with enhancements allowable under SPAR guidelines

Note e - Similar to SPAR enhanced but with additional changes (not typically allowed) to better match plant PRA results

Table B.2 Comparison of SPAR Model Birnbaums with Plant PRA Values (Braidwood 1)

11/18/2003

System	Component Type	Component Description	Plant PRA Birnbaum (1/y)	SPAR Resolution	SPAR Rev. 3	Birnbaum Ratio						
						SPAR Issue PORVs	SPAR Issue DCP	SPAR Issue LOCAs	SPAR Issue HPI	SPAR Issue HRS	SPAR Issue SWS/CCW	SPAR Issue PCS
HRS	MDP	AF Pump 1A	5.00E-04	0.40	1.03	0.55	0.40	0.40	0.22	0.37	0.40	1.04
HRS	Train (MDP)	Aux Feedwater Train A (TM)	4.48E-04	0.42	0.79	0.57	0.42	0.42	0.23	0.28	0.42	1.07
HRS	DDP	AF Pump 1B	1.25E-04	1.09	13.64	6.19	1.90	1.07	0.88	0.60	1.42	1.81
HRS	Train (DDP)	Aux Feedwater Train B (TM)	8.76E-05	1.08	25.11	7.94	2.24	1.06	0.92	0.87	1.55	1.65
RHR	MOV	Charging Pump to Cold Leg Injection isol Valve	5.51E-05	0.82	0.32	0.34	1.96	0.82	0.82	0.64	0.82	0.83
RHR	MDP	RH Pump 1B	5.48E-05	0.94	0.44	0.46	2.10	0.94	0.94	0.77	0.94	0.95
RHR	Train (MDP)	RH Pump 1B (TM)	5.30E-05	0.75	0.24	0.26	1.94	0.75	0.75	0.58	0.75	0.76
CCW	MDP	CC Pump 1A	4.73E-05	0.26	0.21	0.25	0.35	0.14	0.26	0.24	0.26	0.26
RHR	MOV	Charging Pump to Cold Leg Injection isol Valve	3.34E-05	0.54	0.50	0.53	0.54	0.54	0.54	0.49	0.54	0.56
RHR	MDP	RH Pump 1A	3.13E-05	0.78	0.75	0.78	0.79	0.78	0.78	0.73	0.78	0.81
RHR	Train (MDP)	RH Pump 1A (TM)	3.07E-05	0.41	0.39	0.41	0.42	0.41	0.41	0.38	0.41	0.43
HPI	MOV	RH HX B to CV Pump suction isol valve	2.63E-05	1.67	0.62	0.67	4.06	1.66	1.66	1.30	1.67	1.69
HPI	Train (MDP)	SI Pump Train 1B (TM)	2.52E-05	0.01	0.01	0.01	0.01	0.01	0.01	0.01	0.01	0.01
EAC	EDG	EDG 1A	2.51E-05	1.47	1.89	2.46	1.47	1.45	1.46	1.20	1.47	1.46
HPI	MDP	CV Pump 1A	2.00E-05	0.89	0.32	0.44	0.46	0.09	0.45	0.44	0.83	0.45
HPI	MDP	CV Pump 1B	2.00E-05	0.22	0.04	0.44	0.46	0.09	0.45	0.41	0.83	0.45
EAC	Train (EDG)	EDG 1A (TM)	1.97E-05	1.09	1.64	2.26	1.10	1.07	1.09	0.85	1.10	1.09
SWS	MDP	SX Pump 1B	1.88E-05	1.41	1.76	1.41	1.42	1.41	1.41	1.22	2.02	1.41
EAC	EDG	EDG 1B	1.21E-05	2.11	1.88	2.14	2.12	2.11	2.11	1.82	2.16	2.11
CCW	MDP	CC Pump 1B	7.86E-06	0.46	0.19	0.45	0.47	0.21	0.46	0.42	0.46	0.49
SWS	Train (MDP)	SX Pump 1A (TM)	6.20E-06	1.50	1.50	1.45	1.81	0.43	1.49	1.44	2.02	1.55
SWS	Train (MDP)	SX Pump 1B (TM)	6.20E-06	1.50	1.50	1.45	1.81	0.43	1.49	1.44	2.02	1.55
SWS	MDP	SX Pump 1A	5.03E-06	5.58	3.93	5.59	5.61	3.25	5.58	5.18	7.23	5.70
EAC	Train (EDG)	EDG 1B (TM)	4.27E-06	2.60	2.38	2.69	2.61	2.59	2.60	2.24	2.75	2.60
HPI	MOV	CV Pump to Cold Leg injection isol valve	3.10E-06	0.03	0.00	0.00	0.10	0.03	0.03	0.02	0.03	0.03
HPI	MOV	CV Pump to Cold Leg injection isol valve	3.10E-06	0.76	0.07	0.08	0.87	0.76	0.76	0.52	0.76	0.78
HPI	MDP	SI Pump 1A	2.57E-06	0.00	0.00	0.00	0.00	0.00	0.00	0.00	0.00	0.00
HPI	MDP	SI Pump 1B	2.57E-06	0.09	0.08	0.09	0.09	0.08	0.09	0.06	0.09	0.09
HPI	MOV	RH HX A to CV Pump suction isol valve	1.85E-06	9.05	8.41	9.02	9.15	9.04	9.04	8.19	9.04	9.48
HPI	MDP	SI Pump Train 1A (TM)	1.66E-06	0.00	0.00	0.00	0.00	0.00	0.00	0.00	0.00	0.00
RHR	MOV	CC water from RH HX isol Valve	1.54E-06	5.82	5.06	5.79	5.88	5.81	5.81	4.79	5.82	6.35
RHR	MOV	CC water from RH HX isol Valve	1.54E-06	23.46	5.73	6.43	64.57	23.45	23.42	17.22	23.44	23.87
CCW	MOV	Unit 1 CC HX Outlet MOV	1.44E-06	2.21	1.05	1.97	2.83	1.19	2.20	2.02	2.26	2.26
HPI	MOV	VCT Outlet isol Valve	1.39E-06	3.12	0.97	3.06	3.26	0.63	3.12	3.10	5.09	3.12
HPI	MOV	RWST to CV Pump Suction Valve	1.39E-06	4.78	1.13	3.27	5.03	2.26	4.78	4.25	6.75	4.83

Table B.2 Comparison of SPAR Model Birnbaums with Plant PRA Values (Braidwood 1) (continued)

Braidwood 1
11/18/2003

System	Component Type	Component Description	Plant PRA Birnbaum (1/y)	SPAR Resolution	SPAR Rev. 3	Birnbaum Ratio						
						SPAR Issue PORVs	SPAR Issue DCP	SPAR Issue LOCAs	SPAR Issue HPI	SPAR Issue HRS	SPAR Issue SWS/CCW	SPAR Issue PCS
HPI	MOV	VCT Outlet isol Valve	9 12E-07	3 82	1 30	3 73	4 03	0 79	3 82	3 79	6 82	3 82
HPI	MOV	RWST to CV Pump Suction Valve	9 12E-07	3 87	1 31	3 75	4 08	0 82	3 86	3 81	6 86	3 87
HPI	Train (MDP)	CV Pump Train 1B (TM)	9 06E-07	2 53	0 42	1 26	2 99	1 60	2 53	2 09	2 93	2 56
CCW	MOV	Unit 0 CC HX Outlet MOV	9 03E-07	2 43	1 31	2 34	2 66	1 43	2 43	2 23	2 45	2 52
HPI	Train (MDP)	CV Pump Train 1A (TM)	5 69E-07	4 03	0 66	2 01	4 76	2 55	4 02	3 33	4 67	4 08
RHR	MOV	RH HX Discharge Crosstie Valve	3 43E-07									
RHR	MOV	RH HX Discharge Crosstie Valve	3 43E-07									
CCW	Train (MDP)	CC Pump 1A (TM)	2 46E-08	75 03	30 67	73 56	76 36	34 43	74 90	67 95	75 35	79 31
CCW	Train (MDP)	CC Pump 1B (TM)	2 46E-08	75 03	30 67	73 56	76 36	34 43	74 90	67 95	75 35	79 31
		Plant Birnbaum >= 1E-5/y										
		Geometric mean		0 61	0 64	0 68	0 81	0 50	0 56	0 50	0 69	0 72
		Standard deviation		0 53	6 04	2 06	0 98	0 57	0 54	0 43	0 59	0 56
		1E-5/y > Plant Birnbaum >= 1E-6/y										
		Geometric mean		0 67	0 36	0 39	0 81	0 43	0 67	0 56	0 75	0 69
		Standard deviation		5 67	2 43	2 70	15 21	5 77	5 66	4 28	5 72	5 79
		Core Damage Frequency (1/y)	3 01E-05	3 35E-05	9 40E-05	6 09E-05	3 79E-05	2 64E-05	3 17E-05	2 93E-05	6 76E-05	3 98E-05
		Ratio of SPAR CDF to Plant PRA CDF	N/A	1 11	3 12	2 02	1 26	0 88	1 05	0 97	2 25	1 32

Birnbaum ratio is SPAR Birnbaum divided by plant PRA Birnbaum

Table B.3 List of Braidwood 1 SPAR Resolution Model Changes not Allowed Under SPAR Development Guidelines

SPAR Issue Category	Basic Event Affected or Description of Change	SPAR Enhanced Model Value or Description	Change
PORVs	PORV success criterion change for feed-and-bleed	2 of 2 PORVs required for feed-and-bleed	1 of 2 PORVs required for feed-and-bleed
DCP	IE-LDCA	2.4E-7/h	7.3E-8/h
	DCP-BDC-LP-1A	9.0E-5	9.0E-6
	DCP-BDC-LP-1B	9.0E-5	9.0E-6
	DCP-BDC-LP-2A	9.0E-5	9.0E-6
	DCP-BDC-LP-2B	9.0E-5	9.0E-6
LOCAs	RCS-MDP-LK-SEALS	1.9E-1	True
HPI	HPI-XHE-XM-FB2	1.6E-1	5.1E-1
HRS	AFW-MDP-FS-1A	2.8E-3 (*0.21 nonrecovery)	1.6E-3
	AFW-MDP-FR-1A	7.6E-4 (*0.75 nonrecovery)	3.2E-3
	AFW-MDP-TM-1A	1.1E-3	5.2E-3
	AFW-DDP-FS-1B	2.3E-2 (*0.25 nonrecovery)	1.3E-2
	AFW-PMP-CF-ALL	6.2E-8	3.3E-4
	AFW-XHE-XL-MDPFS	2.1E-1	True
	AFW-XHE-XL-MDPFR	7.5E-1	True
	AFW-XHE-XL-EDPFS	2.5E-1	True
	AFW-XHE-XL-EDPFR	7.5E-1	True
SWS/CCW	IE-LOESW	1.1E-7/h	6.0E-9/h
	ESW-MDP-FS-1A	3.0E-3	1.4E-3
	ESW-MDP-FS-1B	3.0E-3	1.4E-3
	ESW-MDP FS 2A	3.0E 3	1.4E-3
	ESW-MDP-FS-2B	3.0E-3	1.4E-3
	ESW-MDP-TM-1A	9.8E-3	5.9E-3
	ESW-MDP-TM-1B	9.8E-3	5.9E-3
	ESW-MDP-TM-2B	9.8E-3	5.9E-3
PCS	MFW-SYS-UNAVAIL	1.0E-1	Ignore
	MFW-XHE-ERROR	1.0E-2	5.3E-3
	PCS-XHE-XO-SEC	2.0E-1	True
	PCS-XHE-XO-SECL	3.4E-1	True

Table B.4 Comparison of SPAR Model MSPI Predictions (4Q2002 Data Set) with Plant PRA Values (Braidwood 1)

MSPI Results for 4th Quarter 2002
Braidwood 1
12/19/2003

System	Plant PRA Model	SPAR Resolution Model	SPAR Rev. 3 Model	SPAR Enhanced Model	SPAR Issue 1/2 PORVs	SPAR Issue DC Power	SPAR Issue LOCAs	SPAR Issue HPI	SPAR Issue HRS	SPAR Issue SWS/CCW	SPAR Issue PCS
EAC	-9.58E-08	-1.57E-07	-9.46E-08	-1.80E-07	-2.26E-07	-1.57E-07	-1.55E-07	-1.56E-07	-1.30E-07	-1.59E-07	-1.56E-07
HPI	4.39E-08	8.50E-08	2.85E-08	4.29E-08	5.27E-08	1.62E-07	7.60E-08	8.44E-08	7.08E-08	9.33E-08	8.63E-08
HRS	2.28E-06	2.58E-06	3.51E-05	3.77E-05	1.57E-05	4.69E-06	2.54E-06	2.13E-06	1.52E-06	3.43E-06	4.17E-06
RHR	1.51E-08	3.95E-11	-1.14E-08	-1.47E-08	-1.57E-08	3.91E-08	2.70E-10	2.83E-10	-3.21E-09	2.94E-10	-5.34E-10
SWS/CCW	6.13E-08	4.09E-08	-1.03E-08	4.97E-08	4.74E-08	1.42E-08	6.03E-08	4.24E-08	3.55E-08	4.81E-08	4.11E-08

MSPI Results for 4th Quarter 2002
Braidwood 2
12/19/2003

System	Plant PRA Model	SPAR Resolution Model	SPAR Rev. 3 Model	SPAR Enhanced Model	SPAR Issue 1/2 PORVs	SPAR Issue DC Power	SPAR Issue LOCAs	SPAR Issue HPI	SPAR Issue HRS	SPAR Issue SWS/CCW	SPAR Issue PCS
EAC	-1.62E-07	-2.49E-07	-1.45E-07	-2.97E-07	-3.79E-07	-2.48E-07	-2.45E-07	-2.47E-07	-2.04E-07	-2.51E-07	-2.47E-07
HPI	-2.00E-08	-1.86E-08	-3.54E-09	-5.15E-09	-1.11E-08	-2.41E-08	-1.13E-08	-1.80E-08	-1.54E-08	-2.27E-08	-1.82E-08
HRS	1.22E-07	3.39E-07	7.05E-06	7.40E-06	2.76E-06	7.42E-07	3.32E-07	3.15E-07	2.02E-07	5.02E-07	4.25E-07
RHR	1.71E-07	1.41E-07	3.93E-08	8.17E-08	8.52E-08	2.73E-07	1.40E-07	1.40E-07	1.18E-07	1.40E-07	1.42E-07
SWS/CCW	6.99E-08	7.88E-08	-9.55E-09	7.05E-08	8.45E-08	5.21E-08	7.99E-08	7.98E-08	6.98E-08	9.10E-08	8.00E-08

Braidwood 2 results use the Braidwood 1 SPAR model importances and CDFs with Braidwood 2 failures, demands and operating hours.

Table B.5 Comparison of SPAR Model MSPI Difference Factors (Braidwood 1)

Braidwood 1
11/18/2003
1 Failure > Baseline
Plant Results

System	Component	Failure Mode	Delta CDF (1/y)	SPAR Resolution	SPAR Rev. 3	SPAR Issue PORVs	SPAR Issue ACP	SPAR Issue DCP	SPAR Issue LOCAs	SPAR Issue HPI	SPAR Issue HR3	SPAR Issue RHR	SPAR Issue SWS/CCW	SPAR Issue PCS	SPAR Issue Misc
								Difference Factor Comparisons							
EAC	EDG	FTS	1.91E-07	0.12	0.17	0.26		0.12	0.12	0.12	0.07		0.13	0.12	
		FTLR	1.66E-07	0.11	0.15	0.22		0.11	0.10	0.11	0.06		0.11	0.11	
		FTR	2.83E-07	0.19	0.25	0.38		0.19	0.18	0.19	0.11		0.19	0.19	
HPI	MDP	FTS	6.89E-08	-0.03	-0.06	-0.03		-0.03	-0.06	-0.03	-0.03		0.00	-0.03	
		FTR	8.71E-09	0.00	-0.01	0.00		0.00	-0.01	0.00	0.00		0.00	0.00	
	MDP Stby	FTS	3.50E-08	-0.03	-0.03	-0.03		-0.03	-0.03	-0.03	-0.03		-0.03	-0.03	
		FTR	3.15E-08	-0.03	-0.03	-0.03		-0.03	-0.03	-0.03	-0.03		-0.03	-0.03	
	MOV	FTO/C	6.04E-08	0.05	-0.01	0.01		0.13	0.03	0.05	0.03		0.06	0.05	
	AOV	FTO/C	1.07E-09	0.00	0.00	0.00		0.00	0.00	0.00	0.00		0.00	0.00	
HRS	MDP Stby	FTS	2.34E-06	-1.39	-0.08	-1.04		-1.39	-1.39	-1.82	-1.53		-1.39	0.11	
		FTR	1.83E-06	-1.09	-0.09	-0.81		-1.09	-1.09	-1.42	-1.20		-1.08	0.09	
	DDP	FTS	1.10E-06	0.09	15.37	5.94		1.03	0.08	-0.12	-0.41		0.48	0.87	
		FTR	1.30E-06	0.11	17.93	6.99		1.21	0.09	-0.14	-0.49		0.56	1.03	
	MOV	FTO/C	3.80E-07	0.18	1.43	0.29		0.11	0.18	-0.25	-0.24		-0.15	0.06	
	AOV	FTO/C	3.80E-07	-0.18	1.43	0.29		-0.11	-0.18	-0.25	-0.24		-0.15	0.06	
RHR	MDP Stby	FTS	3.17E-07	-0.05	-0.16	-0.15		0.18	-0.05	-0.05	-0.09		-0.05	-0.05	
		FTR	2.48E-07	-0.04	-0.13	-0.12		0.14	-0.04	-0.04	-0.08		-0.04	-0.04	
	MOV	FTO/C	1.60E-07	-0.01	-0.09	-0.08		0.18	-0.01	-0.01	-0.04		-0.01	0.00	
3W3	MDP	FTO	3.37E-08	0.04	0.03	0.05		0.06	0.02	0.05	0.04		0.08	0.05	
		FTR	1.25E-08	0.01	0.01	0.02		0.03	0.00	0.02	0.02		0.03	0.02	
CCW	MDP	FTS	6.95E-08	-0.05	-0.05	-0.05		-0.04	-0.06	-0.05	-0.05		-0.05	-0.05	
		FTR	8.55E-09	0.00	-0.01	0.00		0.00	-0.01	0.00	0.00		0.00	0.00	
	MOV	FTO/C	3.25E-09	0.01	0.00	0.01		0.01	0.00	0.01	0.01		0.01	0.01	
Difference factor average				-0.10	1.57	0.53		0.02	-0.11	-0.16	-0.18		-0.06	0.11	
Difference factor standard deviation				0.37	4.79	1.91		0.52	0.37	0.47	0.40		0.41	0.27	

Difference factor = (delta CDF,SPAR - delta CDF, Plant)/(1E-6/y)

Table B.6 Summary of SPAR Model CDFs and Birnbaums with Plant PRA Values

Plant	CDF Comparison CDF(SPAR)/ CDF(Plant PRA)		Birnbaum Comparison (Components with Birnbaum > 1.0E-5/y) Birnbaum(SPAR)/ Birnbaum(Plant PRA) (Geometric Average of Components)		Birnbaum(SPAR)/ Birnbaum(Plant PRA) (Standard Deviation of Components)		Birnbaum Comparison (Components with Birnbaum in range 1.0E-5/y to 1.0E-6/y) Birnbaum(SPAR)/ Birnbaum(Plant PRA) (Geometric Average of Components)		Birnbaum(SPAR)/ Birnbaum(Plant PRA) (Standard Deviation of Components)	
	SPAR Revision 3	SPAR Resolution	SPAR Revision 3	SPAR Resolution	SPAR Revision 3	SPAR Resolution	SPAR Revision 3	SPAR Resolution	SPAR Revision 3	SPAR Resolution
Braidwood 1	3.12	1.11	0.64	0.61	6.04	0.53	0.36	0.67	2.43	5.67
Hope Creek	1.89	1.39	1.12	1.40	0.92	0.12	3.35	1.67	2.98	2.52
Limerick 1	2.22	1.12	3.52	1.49	3.21	1.19	3.18	1.75	3.14	1.09
Millstone 2	0.60	1.30	0.04	1.20	0.90	1.34	0.02	6.87	6.09	Undefined
Millstone 3	2.06	0.86	0.11	0.34	3.29	1.23	0.11	0.34	3.94	0.27
Palo Verde 1	1.28	0.88	0.67	0.78	2.26	0.52	3.21	0.99	4.04	0.66
Prairie Island 1	0.60	0.67	0.60	0.65	0.42	0.15	0.36	0.55	0.76	1.18
Salem 1	1.68	0.97	0.23	0.98	1.32	1.64	0.05	0.45	2.64	3.45
San Onofre 2	1.84	1.34	0.24	0.82	4.52	0.94	0.05	0.73	8.00	1.93
South Texas 1	1.27	1.21	0.12	0.37	0.33	0.69	0.13	0.35	0.19	0.66
Surry 1	1.33	1.44	0.01	5.28	1.60	1.90	1.09	0.61	7.15	2.40
Average	1.63	1.12	0.66	1.27	2.26	0.93	1.08	1.36	3.53	1.98

For the CDF and Birnbaum (geometric average of components) comparisons, a value of 1.00 indicates agreement between the SPAR and plant PRA results. If the value is > 1.00, then the SPAR value is higher than the plant PRA value. For the standard deviation comparisons, a value of 0.00 indicates agreement between the SPAR and plant PRA results.

Table B.7 Summary of SPAR Model MSPI Color Predictions (4Q2002 Data Set) versus Plant PRA Colors

MSPI Color Summary by SPAR Model and System (4Q2002 Data Set)

Plant	SPAR Rev. 3					SPAR Resolution				
	EAC	HPI	HRS	RHR	SWS/CCW	EAC	HPI	HRS	RHR	SWS/CCW
Braidwood 1	G/G	G/G	Y/W	G/G	G/G	G/G	G/G	W/W	G/G	G/G
Braidwood 2	G/G	G/G	W/G	G/G	G/G	G/G	G/G	G/G	G/G	G/G
Hope Creek	G/G	G/G	G/G	G/G	G/G	G/G	G/G	G/G	G/G	G/G
Limerick 1	G/G	G/G	G/G	G/G	G/G	G/G	G/G	G/G	G/G	G/G
Limerick 2	G/G	G/G	G/G	G/G	G/G	G/G	G/G	G/G	G/G	G/G
Millstone 2	G/G	G/G	G/G	G/G	G/G	G/G	G/G	G/G	G/G	G/G
Millstone 3	G/G	G/G	G/G	G/G	G/G	G/G	G/G	G/G	G/G	G/G
Palo Verde 1	G/G	G/G	G/G	G/G	G/G	G/G	G/G	G/G	G/G	G/G
Palo Verde 2	G/G	G/G	G/G	G/G	G/G	G/G	G/G	G/W	G/G	G/G
Palo Verde 3	G/G	G/G	G/G	G/G	G/G	G/G	G/G	G/G	G/G	G/G
Prairie Island 1	C/C	C/C	G/G	G/G	G/G	G/G	G/G	G/G	G/G	G/G
Prairie Island 2	G/G	G/G	G/G	G/G	G/G	G/G	G/G	G/G	G/G	G/G
Salem 1	W/W	G/G	G/G	G/G	G/G	W/W	G/G	G/G	G/G	G/G
Salem 2	G/G	G/G	G/G	G/G	G/G	G/G	G/G	G/G	G/G	G/G
San Onofre 2	G/G	G/G	G/G	G/G	G/G	G/G	G/G	G/G	G/G	G/G
San Onofre 3	G/G	G/G	G/G	G/G	G/G	G/G	G/G	G/G	G/G	G/G
South Texas 1	G/G	G/G	G/G	G/G	G/G	G/G	G/G	G/G	G/G	G/G
South Texas 2	G/G	G/G	G/G	G/G	G/G	G/G	G/G	G/G	G/G	G/G
Surry 1	W/G	G/G	G/G	G/G	G/G	G/G	G/G	G/G	G/G	G/G
Surry 2	W/G	G/G	G/G	G/G	G/G	G/G	G/G	G/G	G/G	G/G

Note - Each entry (e.g., G/W) indicates the system color predicted by the SPAR model and then the color from the plant PRA model. Cases where the colors do not agree are highlighted. Results do not include the frontstop, backstop, or application of CCF multipliers.

Table B.8 Summary of SPAR Model MSPI Difference Factor Predictions

Plant	Difference Factor Comparison (MSPI Delta CDF with 1 Failure Above Baseline)			
	Difference Factor (Arithmetic Average of Component Failure Modes)		Difference Factor (Standard Deviation of Component Failure Modes)	
	SPAR Revision 3	SPAR Resolution	SPAR Revision 3	SPAR Resolution
Braidwood 1	1.57	-0.10	4.79	0.37
Hope Creek	-0.08	0.10	0.54	0.12
Limerick 1	0.10	0.04	0.09	0.06
Millstone 2	-0.67	-0.20	0.95	0.59
Millstone 3	0.53	-0.07	1.21	0.24
Palo Verde 1	-0.09	-0.18	1.09	0.58
Prairie Island 1	-0.06	-0.05	0.09	0.04
Salem 1	-0.21	0.14	0.63	0.53
San Onofre 2	0.42	0.02	0.96	0.25
South Texas 1	-0.09	-0.03	0.08	0.07
Surry 1	0.10	0.03	0.15	0.06
Average	0.14	-0.03	0.96	0.26

For the difference factor (arithmetic average of component failure modes) comparisons, a value of 0.00 indicates agreement between the SPAR and plant PRA results. If the difference factor value is > 0.00, then the SPAR MSPI value is higher than the plant PRA value. For the standard deviation comparisons, a value of 0.00 indicates agreement between the SPAR and plant PRA results.

Table B.9 Summary of SPAR Model MSPI Difference Factor Predictions (Means) for SPAR Issue Categories

Plant	SPAR Rev. 3	SPAR Resolution	SPAR Issue PORVs	SPAR Issue ACP	SPAR Issue DCP	SPAR Issue LOCAs	SPAR Issue HPI	SPAR Issue HRS	SPAR Issue RHR	SPAR Issue SWS/CCW	SPAR Issue PCS	SPAR Issue Misc.
					MSPI Difference Factor Summary (Average of All Monitored Component Failure Modes within a Plant)							
Braidwood 1	1.57	-0.10	0.53		0.02	-0.11	-0.16	-0.18		-0.06	0.11	
Hope Creek	-0.08	0.10		0.40		0.11	-0.05	0.18	0.10	0.25	0.35	0.27
Limerick 1	0.10	0.04		0.05			0.01	0.01	0.04	0.04	0.17	0.05
Millstone 2	-0.67	-0.20		-0.21	-0.19	9.62	-0.27	-0.27		-0.44	-0.20	
Millstone 3	0.53	-0.07		0.16	-0.02	-0.07	-0.18	-0.23	-0.07	-0.07	-0.25	-0.09
Palo Verde 1	-0.09	-0.18			-0.20	-0.16	-0.17	0.05		-0.14		
Prairie Island 1	-0.06	-0.05		-0.07	-0.05	-0.07	-0.05	-0.07	-0.04	-0.02		-0.06
Salem 1	-0.21	0.14		-0.13		0.10	0.14			0.72		0.15
San Onofre 2	0.42	0.02		0.04		-0.05	0.02	0.00		0.00		0.02
South Texas 1	-0.09	-0.03		-0.02		-0.03	-0.03	-0.03			-0.03	
Surry 1	0.10	0.03		0.07	0.04	0.02		0.04		0.03		

The difference factor is defined as (delta CDF, SPAR - delta CDF,plant PRA)/1.0E-6/y. Results for each plant represent averages of the difference factors calculated for each of the monitored component failure modes.

SPAR issue values highlighted in grey indicate cases where the issue results in an average difference factor that is +/- 0.50 worse than the SPAR resolution result.

A difference factor of 0.00 represents agreement between the SPAR and plant PRA results. If the value is > 0.00, then the SPAR delta CDF prediction is higher than the plant PRA prediction. If the value is < 0.00, then the SPAR delta CDF prediction is lower.

Table B.10 Summary of SPAR Model MSPI Difference Factor Predictions (Standard Deviations) for SPAR Issue Categories

MSPI Difference Factor Summary (Standard Deviation of All Monitored Component Failure Modes within a Plant)

Plant	SPAR Rev. 3	SPAR Resolution	SPAR Issue PORVs	SPAR Issue ACP	SPAR Issue DCP	SPAR Issue LOCAs	SPAR Issue HPI	SPAR Issue HRS	SPAR Issue RHR	SPAR Issue SWS/CCW	SPAR Issue PCS	SPAR Issue Misc.
Braidwood 1	4.79	0.37	1.91		0.52	0.37	0.47	0.40		0.41	0.27	
Hope Creek	0.54	0.12		0.47		0.16	0.60	0.19	0.12	0.37	0.54	0.35
Limerick 1	0.09	0.06		0.06		0.04	0.04	0.04	0.06	0.06	0.25	0.07
Millstone 2	0.95	0.59		0.57	0.60	16.77	0.63	0.57		0.76	0.59	
Millstone 3	1.21	0.24		0.35	0.35	0.25	0.21	0.28	0.24	0.24	0.20	0.23
Palo Verde 1	1.09	0.58			0.59	0.58	0.58	0.79		0.58		
Prairie Island 1	0.09	0.04		0.06	0.04	0.05	0.05	0.08	0.05	0.07		0.05
Salem 1	0.63	0.53		0.58		0.55	0.53			1.85		0.52
San Onofre 2	0.96	0.25		0.28		0.30	0.25	0.23		0.26		0.25
South Texas 1	0.08	0.07		0.09		0.07	0.06	0.07			0.07	
Surry 1	0.15	0.06		0.14	0.11	0.09		0.11		0.07		

The difference factor is defined as (delta CDF, SPAR - delta CDF,plant PRA)/1.0E-6/y. Results for each plant represent difference factor standard deviations calculated from results for each of the monitored component failure modes.

SPAR issue values highlighted in grey indicate cases where the issue results in a difference factor standard deviation that is 0.50 worse than the SPAR resolution result.

A difference factor standard deviation of 0.00 indicates agreement between the SPAR and plant PRA results. The standard deviation can only be >= 0.00. A higher value indicates a poorer match of SPAR delta CDF predictions with plant PRA predictions.

Table B.11 Summary of SPAR Model Issue Category Impacts on MSPI Predictions

SPAR Model Issue Category Impact on MSPI Prediction (1 Failure Above Baseline)

Potential Impact on MSPI Prediction	SPAR Issue PORVs	SPAR Issue ACP	SPAR Issue DCP	SPAR Issue LOCAs	SPAR Issue HPI	SPAR Issue HRS	SPAR Issue RHR	SPAR Issue SWS/CCW	SPAR Issue PCS	SPAR Issue Misc.
Large (>5.0E-7/y)	Braidwood			Millstone 2				Salem		
Medium (1.0E-7/y to 5.0E-7/y)		Hope Creek Millstone 3 Salem	Braidwood		Hope Creek Millstone 3	Millstone 3 Palo Verde		Hope Creek Millstone 2	Braidwood Hope Creek Limerick Millstone 3	Hope Creek
Small (<1.0E-7/y)	All others	All others	All others	All others	All others	All others	All	All others	All others	All others

B-21

APPENDIX C. TECHNICAL BASIS FOR REVISED BASELINE COMPONENT FAILURE RATES

Appendix C
Technical Basis for Revised
Baseline Component Failure Rates

C.1   Summary

The Mitigating Systems Performance Index (MSPI) pilot program investigated whether component performance during the period 1995 – 1997 is significantly different from performance during the period 1999 – 2001, and whether or not performance data from 1999 – 2001 should be used in lieu of Table 2 baselines of NEI 99-02 (shown below in Table C.1). To investigate these issues, two data sources were reviewed: Equipment Performance and Information Exchange (EPIX), and Licensee Event Reports (LERs) used in the updated system studies. Statistical trend analyses of each of these data sources indicate no significant trends over the period 1995 – 2001, except for the auxiliary feedwater system diesel-driven pump failure to run (FTR) rate, which has an increasing rate with time.

Ignoring the statistical evidence of essentially no trends, if component failure mode data are fitted to trend curves, then the geometric average of the ratios of 1996 estimate to 2000 estimate for the component failure modes is 1.25 using the EPIX data and 1.18 using the updated system study (LER) data. This composite result indicates that 1996 performance may be approximately 18% to 25% worse than 2000 performance. Therefore, this composite metric also indicates little difference between 1996 and 2000 component performance.

The Year 2000 baselines proposed by the Nuclear Regulatory Commission (NRC) for use in the MSPI pilot program (Table C.2) appear to be approximately 16% high (overall for the 16 component failure modes used in the MSPI) when compared with actual pilot plant performance for the period July 1999 through June 2002. The apparent 16% higher values compensate for most of the potential 18 to 25% difference between 1996 and 2000 performance. The fact that all these performances are so close also gives reasonable confidence in the appropriateness of the revised Year 2000 failure rates.

The existing Table 2 baselines in draft NEI 99-02 are not representative of component performance for the period 1995 – 1997. The sources used to develop the existing Table 2 are more representative of component performance around 1990 or 1991 (or earlier).

Therefore, the Year 2000 baselines are recommended for use in the MSPI pilot program because of the following:

- The existing Table 2 baselines are representative of component performance around 1990 or 1991 (or earlier), not for the period 1995 – 1997.

- There appears to be little or no trend in component performance over the period 1995 – 2001.

- The Year 2000 baselines were all generated using a single consistent set of industry data matching the types of data to be reported in the MSPI pilot program.

- The Year 2000 baselines appear to be 16% high compared with current performance. Therefore, the apparent 16% higher values compensate for most of the potential 18 to 25 percent difference between 1996 and 2000 performance.

- Using Year 2000 baselines is consistent with the MSPI train unavailability baselines, which were also generated from data for the period 1999 – 2001.

C.2   Introduction

Component baseline failure rates for the MSPI pilot program are presented in Table 2 in Appendix F of the draft NEI 99-02 report. Those baseline failure rates were generated by the NRC in early 2002. At that time, the most appropriate published sources for component baseline failure rates were judged to be the system studies (NUREG/CR-5500 series) published in the late 1990s and the generic database developed for the NUREG-1150 studies (NUREG/CR-4550, Vol. 1). The desire was to generate component baseline failure rates representative of industry performance over the period 1995 – 1997. However, the available published sources were more representative of industry performance around approximately 1990 or 1991. Therefore, the existing Table 2 values are representative of industry performance around 1990 or 1991, and not for the period 1995 – 1997.

Follow-on work in support of the MSPI pilot program included an update to the Table 2 component baseline failure rates, to reflect industry performance during the period 1999 – 2001. The source for the updated Table 2 values, termed the Year 2000 baselines, was primarily the journal article given in Ref. C.1. Baseline failure rates in that journal article for the period 1999 – 2001 were obtained from the EPIX database maintained by the Institute for Nuclear Power Operations (INPO). The EPIX data were reviewed and evaluated using the Reliability and Availability Database System (RADS) software developed by the NRC.

To more fully complete the component baseline failure rate work for the MSPI pilot program, equipment performance over the period 1995 through 2001 would need to be investigated, to discern whether significant differences exist between the period 1995 – 1997 and 1999 – 2001. If significant differences do exist, then a new set of baselines should be established for the period 1995 – 1997. A decision would then be made whether to use the 1995 – 1997 baselines or the Year 2000 baselines. However, the use of 1999 – 2001 data has been judged to be acceptable by all stakeholders and is the recommended baseline data to implement in the MSPI at this time.

C.3   Existing Table 2 Baselines

The existing Table 2 component baseline failure rates are presented in Table C.1. Several issues need to be kept in mind when reviewing the existing Table 2 mean failure probabilities and rates:

- The failure to start (FTS) probabilities, except for the emergency diesel generators (EDGs), include failures to run that occur within the first hour of operation. This "expanded" definition of FTS was recommended by the NRC to help reduce the number of FTR events. (Such events typically have a greater chance of resulting in a change in core damage frequency greater than 1.0×10^{-6}/year, given just one failure.) Also, this approach is generally consistent with the approach used in the NRC component studies (NUREG-1715 series). To identify such events within the system studies, the individual failure reports were reviewed to determine which FTR events were placed into the FTS category and which remained in the FTR category. Note that this effort was time intensive.

- FTR rates apply only after the first hour of operation.

- Failure probabilities and rates reflect nonrecovery probabilities identified in the system studies. For example, the EDG FTS probability in the existing Table 2 is the product of FTS and failure to recover from FTS. The nonrecovery probabilities range from 0.88 to 0.033, with an average of approximately 0.5. Nonrecovery probabilities were included when generating the existing Table 2 baselines for two reasons: it was not clear at that time whether all failures would be reported (or just those that were not recovered), and it was judged that the system study results were too high if nonrecovery was not included, compared with more recent industry performance. Note that the MSPI guidelines for data reporting instruct the plants to report all failures, not just those that could not be recovered within minutes from the control room (without any actual repair activities).

- For several of the component failure modes [motor-operated valve (MOV) failure to open or close (FTO/C), motor-driver pump (MDP) standby failure to start (FTS), MDP standby failure to run (FTR), TDP standby high-pressure coolant injection (HPCI)/reactor core isolation cooling (RCIC) FTS, and TDP standby FTR], data from several different system studies were combined to obtain the values in the existing Table 2.

- The component boundaries in the system studies are generally broader than those in the MSPI. In August 2002 the system study failure events used to generate the existing Table 2 values were reviewed to identify events outside the component boundaries specified in draft NEI 99-02. (Typically, up to 20% of the FTS events were eliminated, while up to 100% of the FTR events were eliminated, depending upon the component.) These changes were never incorporated into the existing Table 2, mainly because the focus turned to development of the Year 2000 baselines.

- Component baseline failure probabilities and rates are representative of industry performance around 1990 or 1991 (or earlier, in some cases).

- Eleven of sixteen component failure mode baselines were derived from the system studies, while five were obtained from the older NUREG-1150 generic database.

- Given the mean failure probabilities and rates, "a" and "b" parameters for beta or gamma prior distributions are generated assuming a constrained noninformative prior.

C.4   Year 2000 Baselines

As explained in the introduction, the Year 2000 baseline failure probabilities and rates (see Table C.2) were obtained from the journal article given in Ref. C.1. Baseline failure rates in Ref. C.1 for the period 1999 – 2001 were obtained from the EPIX database maintained by INPO. In general, the recommended values in Table 2 of Ref. C.1 were used directly. However, for pump FTS including the first hour of operation, the FTS value in Ref. C.1 was combined with the FTR rate specific for the first hour of operation (multiplied by one hour) to obtain the FTS value for the Year 2000 baselines. (Ref. C.1 subdivided FTR for standby components into two periods: the first hour of operation, and operation beyond the first hour of operation. In general, a factor of approximately 15 difference was observed between the two failure rates, with the first hour of operation having the higher FTR value.) Also, the FTR rates (following the first hour of operation) in Ref. C.1 could then be used directly for Year 2000 FTR baselines. Ref. C.1 did not cover circuit breakers. A separate EPIX/RADS search was performed to determine the mean failure probability for this component. Therefore, all of the component failure mode Year 2000 baselines were generated from the same data source using the same methodology.

Table C.1 Existing Table 2 of NEI 99-02 Component Baseline Failure Rates and Sources

Component	Failure Mode	Applicable MSPI Systems	Mean Failure Probability or Rate	Source	Data Period	Midpoint of Data Period
MOV	FTO/C	All	2.1E-3/d	NUREG/CR-5500, Vol. 4,7,8,9	1987 – 1997	1992
AOV	FTO/C	All	2.0E-3/d	NUREG/CR-4550, Vol. 1	1970 – 1983	1977
MDP Standby	FTS [a]	HPI, HPCS, AFW, RHR, SWS, CCW	2.1E-3/d	NUREG/CR-5500, Vol. 1,8,9	1987 – 1995	1991
	FTR [b]	HPI, HPCS, AFW, RHR, SWS, CCW	1.0E-4/h	NUREG/CR-5500, Vol. 1,8,9	1987 – 1995	1991
MDP Running or Alternating	FTS [a]	HPI (CVCS), SWS, CCW	3.0E-3/d	NUREG/CR-4550, Vol. 1	1970 – 1983	1977
	FTR [b]	HPI (CVCS), SWS, CCW	3.0E-5/h	NUREG/CR-4550, Vol. 1	1970 – 1983	1977
TDP Standby, AFW	FTS [a]	AFW	1.9E-2/d	NUREG/CR-5500, Vol. 1	1987 – 1995	1991
	FTR [b]	AFW	1.6E-3/h	NUREG/CR-5500, Vol. 1,4,7	1987 – 1995	1991
TDP Standby, HPCI/RCIC	FTS [a]	HPCI, RCIC	2.7E-2/d	NUREG/CR-5500, Vol. 4,7	1987 – 1993	1990
	FTR [b]	HPCI, RCIC	1.6E-3/h	NUREG/CR-5500, Vol. 1,4,7	1987 – 1995	1991
DDP Standby	FTS [a]	AFW, SWS	1.9E-2/d	NUREG/CR-5500, Vol. 1	1987 – 1995	1991
	FTR [b]	AFW, SWS	8.0E-4/h	NUREG/CR-4550, Vol. 1	1970 – 1983	1977
EDG Standby	FTS	EAC	1.1E-2/d	NUREG/CR-5500, Vol. 5	1987 – 1993	1990
	FTLR [c]	EAC	1.7E-3/d [c]	NUREG/CR-5500, Vol. 5	1987 – 1993	1990
	FTR [b]	EAC	2.3E-4/h	NUREG/CR-5500, Vol. 5	1987 – 1993	1990
Circuit Breaker	FTO/C	EAC	3.0E-3/d	NUREG/CR-4550, Vol. 1	1970 – 1983	1977

Acronyms: AFW (auxiliary feedwater system), AOV (air-operated valve), CCW (component cooling water system), CVCS (chemical and volume control system), DDP (diesel-driven pump), EAC (emergency AC power system), EDG (emergency diesel generator), FTLR (fail to load and run for 1h), FTO/C (fail to open or close), FTR (fail to run), FTS (fail to start), HPCI (high-pressure coolant injection system), HPCS (high-pressure core spray), HPI (high-pressure injection system), RCIC (reactor core isolation cooling system), RHR (residual heat removal system), SWS (service water system).

Notes:
a. FTS includes FTR events that occur within the first hour of operation.
b. FTR applies to continued operation after successful start and operation for the first hour.
c. The system study did not address the FTLR failure mode. A value was obtained by multiplying the FTR rate for 0 to 0.5h by 0.5h, doing the same to the FTR rate for 0.5 to 14h, and adding the two results (to cover 1h of operation). This approximation probably underestimates the FTLR probability, while overestimating the FTS probability.

Since Ref. C.1 was published, the Year 2000 baselines have been compared with results from the MSPI pilot plant data submittals (July 1999 through June 2002). That comparison is presented in Table C.3. In general, the agreement is good, keeping in mind that the MSPI pilot program includes only 20 plants, compared with 103 plants in the EPIX database. The existing Table 2 values are also listed in Table C.3 for comparison purposes.

Another related comparison was made using the pilot plant data. Over the period July 1999 – June 2002, the pilot plants experienced 72 failures in components covered by the MSPI pilot program. Using the reported demands and hours, the Year 2000 baselines predict 83.5 failures. Therefore, overall, the Year 2000 baselines appear to be high by approximately 16%, compared with the actual pilot plant performance.

In contrast to the Year 2000 expected failures (83.5), the expected number of failures using the existing Table 2 baselines is 176.9, compared with the actual number of failures, 72. Therefore, the Year 2000 baselines are much closer to pilot plant performance than are the existing Table 2 baselines.

Table C.2 Year 2000 Component Baseline Failure Rates and Sources

Component	Failure Mode	Applicable MSPI Systems	Mean Failure Probability or Rate	Constrained Noninformative Prior Parameters		Source	Data Period	Midpoint of Data Period
				a	b			
MOV	FTO/C	All	7.0E-4/d	0.499	712.0	EPIX/RADS	1999 – 2001	2000
AOV	FTO/C	All	1.0E-3/d	0.498	498.0	EPIX/RADS	1999 – 2001	2000
MDP Standby	FTS[a]	HPI, HPCS, AFW, RHR, SWS, CCW	1.9E-3/d	0.497	261.0	EPIX/RADS	1999 – 2001	2000
	FTR[b]	HPI, HPCS, AFW, RHR, SWS, CCW	5.0E-5/h	0.500	10000.0	EPIX/RADS	1999 – 2001	2000
MDP Running or Alternating	FTS[a]	HPI (CVCS), SWS, CCW	1.0E-3/d	0.498	498.0	EPIX/RADS	1999 – 2001	2000
	FTR[b]	HPI (CVCS), SWS, CCW	5.0E-6/h	0.500	100000.0	EPIX/RADS	1999 – 2001	2000
TDP Standby, AFW	FTS[a]	AFW	9.0E-3/d	0.485	53.3	EPIX/RADS	1999 – 2001	2000
	FTR[b]	AFW	2.0E-4/h	0.500	2500.0	EPIX/RADS	1999 – 2001	2000
TDP Standby, HPCI/RCIC	FTS[a]	HPCI, RCIC	1.3E-2/d	0.478	36.3	EPIX/RADS	1999 – 2001	2000
	FTR[b]	HPCI, RCIC	2.0E-4/h	0.500	2500.0	EPIX/RADS	1999 – 2001	2000
DDP Standby	FTS[a]	AFW, SWS	1.2E-2/d	0.480	39.5	EPIX/RADS	1999 – 2001	2000
	FTR[b]	AFW, SWS	2.0E-4/h	0.500	2500.0	EPIX/RADS	1999 – 2001	2000
EDG Standby	FTS	EAC	5.0E-3/d	0.492	97.9	EPIX/RADS	1999 – 2001	2000
	FTLR	EAC	3.0E-3/d	0.495	164.0	EPIX/RADS	1999 – 2001	2000
	FTR[b]	EAC	8.0E-4/h	0.500	625.0	EPIX/RADS	1999 – 2001	2000
Circuit Breaker	FTO/C	EAC	8.0E-4/d	0.499	623.0	EPIX/RADS	1999 – 2001	2000

a. FTS includes FTR events that occur within the first hour of operation.
b. FTR applies to continued operation after successful start and operation for the first hour.

Table C.3 Year 2000 Baseline Comparison with Pilot Plant Data

Component	Failure Mode	Applicable MSPI Systems	Year 2000 Mean Failure Probability or Rate	Pilot Plant Data Mean Failure Probability or Rate (3Q1999 – 2Q2002)	Existing Table 2 Mean Failure Probability or Rate
MOV	FTO/C	All	7.0E-4/d	1.4E-3/d	2.1E-3/d
AOV	FTO/C	All	1.0E-3/d	6.3E-4/d	2.0E-3/d
MDP Standby	FTS [a]	HPI, HPCS, AFW, RHR, SWS, CCW	1.9E-3/d	4.4E-4/d	2.1E-3/d
	FTR [b]	HPI, HPCS, AFW, RHR, SWS, CCW	5.0E-5/h	3.0E-5/h	1.0E-4/h
MDP Running or Alternating	FTS [a]	HPI (CVCS), SWS, CCW	1.0E-3/d	5.0E-4/d	3.0E-3/d
	FTR [b]	HPI (CVCS), SWS, CCW	5.0E-6/h	1.2E-5/h	3.0E-5/h
TDP Standby, AFW	FTS [a]	AFW	9.0E-3/d	2.8E-3/d	1.9E-2/d
	FTR [b]	AFW	2.0E-4/h	<2.4E-4/h [c]	1.6E-3/h
TDP Standby, HPCI/RCIC	FTS [a]	HPCI, RCIC	1.3E-2/d	<4.7E-3/d [c]	2.7E-2/d
	FTR [b]	HPCI, RCIC	2.0E-4/h	<2.4E-4/h [c]	1.6E-3/h
DDP Standby	FTS [a]	AFW, SWS	1.2E-2/d	2.1E-2/d	1.9E-2/d
	FTR [b]	AFW, SWS	2.0E-4/h	4.8E-3/h	8.0E-4/h
EDG Standby	FTS	EAC	5.0E-3/d	3.1E-3/d	1.1E-2/d
	FTLR	EAC	3.0E-3/d	3.4E-3/d	1.7E-3/d
	FTR [b]	EAC	8.0E-4/h	5.8E-4/h	2.3E-4/h
Circuit Breaker	FTO/C	EAC	8.0E-4/d	No data	3.0E-3/d

a. FTS includes FTR events that occur within the first hour of operation.
b. FTR applies to continued operation after successful start and operation for the first hour.
c. No failures and limited demands or hours, so this value probably overestimates the actual failure probability or rate.

C.5   Equipment Performance Trends over the Period 1995 – 2001

There are two main sources of data available to the NRC that can be used to investigate equipment performance trends over the period 1995 – 2001: EPIX data, and LERs used in the system studies. Both sources of data have shortcomings for this effort. For example, the EPIX data cover the period from 1997 through the present. EPIX data are not available for 1995 and 1996. Also, the LERs cover mainly component failures occurring during unplanned demands (and cyclic tests performed approximately every 18 months), while the MSPI pilot program focuses heavily on failures during monthly or quarterly testing. Both types of data are analyzed in this section.

The available EPIX data for the period 1997 – 2002 were analyzed for trends with time using the RADS software. The analysis included a test for whether a trend actually exists (p-value determination) and the fitting of the yearly data to a curve. For demand-related failures, the RADS curve fit is of the form:

$$\frac{P}{1-P} = e^{X+Yt}$$

where P = component failure mode probability for a given year
X = constant
Y = constant
t = integer representing the year, with 1997 represented by 1, 1998 by 2, etc.

For failures to run, the RADS curve fit is of the form:

$$\lambda = e^{X+Yt}$$

where λ = component failure to run rate (1/h) for a given year.

The EPIX data were not reviewed to eliminate failures outside the component boundaries specified in draft NEI 99-02. Also, FTR data were not reviewed in detail to segregate FTR (<1h) from FTR (>1h). (This effort would be resource intensive.)

Shown in Table C.4 are the p-values from the trend analyses. The smaller the p-value, the more certain the analysis is that there is a trend in P (or λ) with time. Typically in statistical analyses, a p-value of less than 0.05 is used to declare that there is a significant trend with time. With larger p-values, the data are typically processed using a no-trend (homogeneous data) assumption to generate component failure probabilities. (This approach was used in the NRC system studies and initiating event studies.) Using the p-value < 0.05 criterion to declare a trend with time, only the DDP FTS has a trend. The other fifteen component failure modes have no trends. Even using a more relaxed criterion of p-value < 0.20, only four of sixteen component failure modes have trends. An example trend plot from RADS is presented in Figure C.1.

The other type of trend comparison was to use the trend analyses to compare curve fit values for 1996 (midpoint of the period 1995 – 1997) with those for 2000 (midpoint of the period from 1999 through 2001). To extrapolate a value for 1996, t was set to 0 in the trend equations. The ratios P_{1996}/P_{2000} (for FTO/C, FTS, and FTLR) and $\lambda_{1996}/\lambda_{2000}$ (for FTR) are summarized in Table C.4. The ratios range from a high of 2.02 to a low of 0.36, and the geometric average is 1.25. Again, this composite metric indicates only a potentially small change (25%) in component performance between 1996 and 2000.

Updated system study (NUREG/CR-5500 series) data are available for the period 1987 – 2001. These updated studies cover AFW, HPI, HPCI, HPCS, RCIC, and IC (isolation condenser). Note that the earlier system studies also included EAC (EDGs), but that study (1987 – 1993 data) has not been updated through 2001 because plants no longer report EDG data under RG 1.108. Therefore, the updated system studies do not cover EAC (EDGs), RHR, SWS, or CCW.

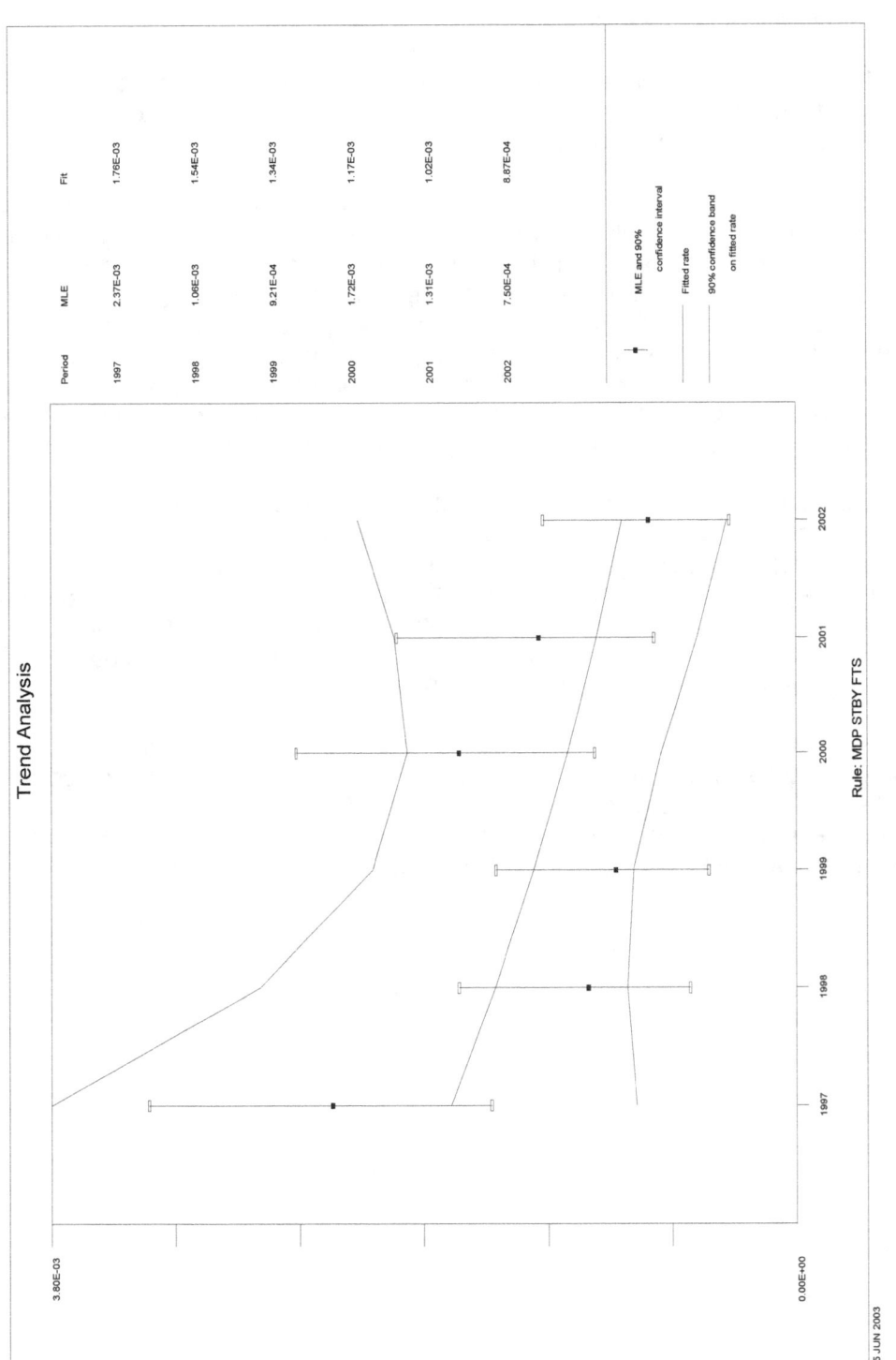

Figure C.1 EPIX MDP (standby) FTS Data Trend Plot (p-value = 0.24)

C-8

Table C.4 EPIX Data Trend Analysis and Comparison (1996 vs. 2000)

Component	Failure Mode	Applicable MSPI Systems	Ratio of Mean Failure Probability or Rate (1996/2000)	Trend Analysis P-Value	Trend Exists? (p-value < 0.05)
MOV	FTO/C	All	2.02	0.07	No
AOV	FTO/C	All	1.74	0.23	No
MDP Standby	FTS	HPI, HPCS, AFW, RHR, SWS, CCW	1.51	0.24	No
	FTR	HPI, HPCS, AFW, RHR, SWS, CCW	1.64	0.33	No
MDP Running or Alternating	FTS	HPI (CVCS), SWS, CCW	1.22	0.53	No
	FTR	HPI (CVCS), SWS, CCW	1.47	0.13	No
TDP Standby, AFW	FTS	AFW	1.96	0.19	No
	FTR	AFW	1.34	0.69	No
TDP Standby, HPCI/RCIC	FTS	HPCI, RCIC	1.76	0.34	No
	FTR	HPCI, RCIC	1.34	0.69	No
DDP Standby	FTS	AFW, SWS	0.36	0.03	Yes
	FTR	AFW, SWS	0.99	0.40	No
EDG Standby	FTS	EAC	1.19	0.51	No
	FTLR	EAC	1.18	0.91 or 0.45	No
	FTR	EAC	0.57	0.48	No
Circuit Breaker	FTO/C	EAC	1.40	0.45	No
Summary		Geometric Average	1.25		15 of 16 component failure modes have no significant trend

The updated system study data were combined across studies similar to the process used to generate the existing Table 2 baselines However, the data were not reviewed to identify FTR events that occurred within the first hour of operation (to place such events into FTS), and to identify failures outside the MSPI component boundaries. (This effort would be resource intensive.)

The combined system study data were plotted versus year, and the results were analyzed for trends. An example trend plot is presented in Figure C.2. Shown in Table C.5 are the p-values from the trend analyses. Using the p-value < 0.05 criterion to declare a trend with time, none of the nine component failure modes covered by the system studies have a significant trend Using the more relaxed criterion of p-value < 0.20, one of nine component failure modes has a trend. However, even if there is no trend for any component, we expect 20% of the data sets (that is, 1.8 of the 9 data sets) to show a p-value < 0.20. So the overall body of data is consistent with the hypothesis of no trend.

The other type of trend comparison made was to use the trend analyses to compare curve fit values for 1996 with those for 2000. The ratios P_{1996}/P_{2000} (for FTO/C, FTS, and FTLR) and $\lambda_{1996}/\lambda_{2000}$ (for FTR) are summarized in Table C.5. The ratios range from a high of 2.04 to a low of 0.95, and the geometric average is 1 18. Again, this composite metric indicates only a potentially small change (18%) in component performance between 1996 and 2000.

C-9

C.6 Reference

C.1   S.A. Eide, "Historical Perspective on Failure Rates for US Commercial Reactor Components," *Reliability Engineering and System Safety,* Vol. 80, 2003, pp. 123 - 132.

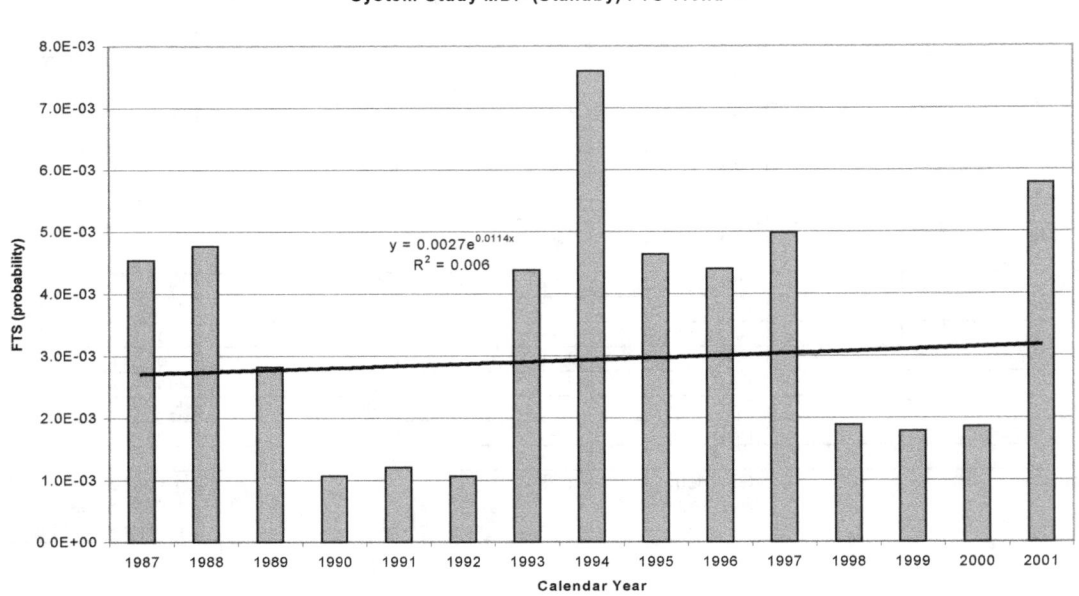

Figure C.2 System Study MDP (standby) FTS Data Trend Plot (p-value = 0.82)

Table C.5 Updated System Study Data Trend Analysis and Comparison (1996 vs. 2000)

Component	Failure Mode	Applicable MSPI Systems	Ratio of Mean Failure Probability or Rate (1996/2000)	Trend Analysis P-Value	Trend Exists? (p-value < 0.05)
MOV	FTO/C	All	0.95	Too few failures to detect a trend	No
AOV	FTO/C	All	No data		
MDP Standby	FTS	HPI, HPCS, AFW, RHR, SWS, CCW	0.96	0.82	No
	FTR	HPI, HPCS, AFW, RHR, SWS, CCW	0.95	Too few failures to detect a trend	No
MDP Running or Alternating	FTS	HPI (CVCS), SWS, CCW	No data		
	FTR	HPI (CVCS), SWS, CCW	No data		
TDP Standby, AFW	FTS	AFW	1.11	0.55	No
	FTR	AFW	1.40	0.39	No
TDP Standby, HPCI/RCIC	FTS	HPCI, RCIC	2.04	0.12	No
	FTR	HPCI, RCIC	1.40	0.39	No
DDP Standby	FTS	AFW, SWS	0.95	Too few failures to detect a trend	No
	FTR	AFW, SWS	0.95	Too few failures to detect a trend	No
EDG Standby	FTS	EAC	No data		
	FTLR	EAC	No data		
	FTR	EAC	No data		
Circuit Breaker	FTO/C	EAC	No data		
Summary		Geometric Average	1.18		9 of 9 component failure modes have no significant trend

APPENDIX D. TECHNICAL BASIS FOR THE FRONTSTOP TO ADDRESS INVALID INDICATORS

Appendix D
Technical Basis for the Frontstop to
Address Invalid Indicators

D.1 Introduction

Some indicators associated with the Mitigating Systems Performance Index (MSPI) proposed by the Nuclear Energy Institute (NEI) (Ref. D.1) have significant "false positive" issues. That is, for statistical reasons, there is a significant probability of a plant system at baseline performance crossing over the GREEN/WHITE threshold. Within the MSPI pilot program, these indicators have been called "overly sensitive indicators" or, in the extreme case, "Invalid Indicators." These designations were given because a small performance change induces a relatively large change in core damage frequency (CDF) and, in the extreme case, this oversensitivity prevents the indicator from being effective. This appendix provides a proposed solution for addressing this issue through the use of a "frontstop."

D.1.1 Frontstop Concept

As defined within the context of the MSPI pilot program, a frontstop is a supplementary set of requirements or adjustments that must be satisfied prior to assigning a "WHITE." These adjustments are designed to ameliorate the indicator's sensitivity, a sensitivity that is, in part, due to the basic simplified approach of the MSPI framework.

A frontstop could be a minimum number of failures, or a fixed or variable risk threshold. Adjustments to the input parameters such as limiting the risk contribution associated with failures could also be used to accomplish the same result. Limiting the risk contribution associated with failures is the approach used for the proposed frontstop.

D.1.2 Sensitive Indicator Issue

Sensitive indicators have a significant probability of exceeding a performance threshold as a result of statistical fluctuations, even if performance is at baseline. An extreme example is an indicator that crosses a threshold as a result of a single failure within an observation period. Ref. D.1 states:

> The performance index relies on the existing testing programs as the source of the data that is input to the calculations. Thus, the number of demands in the monitoring period is based on the frequency of testing required by the current test programs. In most cases, this will provide a sufficient number of demands to result in a valid statistical result. However, in some cases, the number of demands will be insufficient to resolve the change in the performance index (1.0×10^{-6}) that corresponds to movement from a green performance to a white performance level. In these cases, one failure is the difference between baseline performance and performance in the white performance band. The performance index is not suitable for monitoring such systems and monitoring is performed through the inspection process.

The NEI guidance refers to indicators that cross a threshold on a single failure as "Invalid Indicators." There are also valid sensitive indicators; indicators that maintain acceptable performance for all single failures but cross a performance threshold as a result of what could be referred to as expected performance variations.

The issue with both sensitive and invalid indicators is "false positives." Random failures that occur at a rate consistent with the industry performance are not indicative of a performance issue. However, due to limitations associated with the MSPI framework, these random failures could result in challenging the WHITE threshold. These limitations are primarily associated with the data collection duration and the data update process. The collection duration and data update process were designed to achieve an indicator that would minimize the failure to detect degraded performance (false negatives). This balance between preventing the failure to identify degraded performance while not falsely identifying performance as degraded is the driver behind the frontstop. The MSPI is sensitive to changes in equipment reliability to minimize false negatives *to the extent possible* and, therefore, requires some adjustment to prevent false positives.

D.1.3 MSPI Equation

The MSPI is the sum of the unreliability index (URI) and the unavailability index (UAI). The sensitive indicator issue is mainly focused on the URI. However, it is important to understand how the UAI index contributes to ensure that the design of the frontstop works with both indices.

D.1.3.1   MSPI System Unreliability Index (URI)

Equation 3 of Ref. D.1 defines the System Unreliability Index (URI) and is reproduced below. This equation is examined in this appendix since its structure is key to the design of the frontstop.

$$URI = CDF_p \sum_{j=1}^{m} \left[\frac{FV_{URcj}}{UR_{pcj}} \right]_{max} (UR_{Bcj} - UR_{BLcj})$$ (NEI 99-02 Eq. 3)

where the summation is over the number of active components (*m*) in the system, and:

CDF_p is the plant-specific internal events, at power, core damage frequency,

FV_{Urc} is the component-specific Fussell-Vesely value for unreliability,

UR_{Pc} is the plant-specific PRA value of component unreliability,

UR_{Bc} is the Bayesian corrected component unreliability for the previous 12 quarters, and

UR_{BLc} is the historical industry baseline calculated from unreliability mean values for each monitored component in the system.

By examining Equation 3, it can be seen that there are three factors that contribute to the sensitivity of performance indices:

- sensitivity to changes in Unreliability (ΔUR)
- high Fussell-Vesely importance (FV/UR)
- high CDF

Each of these issues is examined below.

(Note that the product of CDF*(FV/UR) is equivalent to the Birnbaum Importance Measure; this measure is referred to later in this appendix and is described in Ref. D.2.)

D.1.3.1.1   Sensitivity to Changes in Unreliability (ΔUR)

For highly reliable components, a single failure can cause a large change in unreliability. Several solutions have been investigated in an attempt to reduce or eliminate this sensitivity. These include pooling data, merging failure modes, and modifying the data update process. Improvements in the pooling of data and the treatment of failure modes have been incorporated into the MSPI framework. Although the proposed frontstop benefits from these improvements it does not directly include these elements in its design. The solutions to sensitive data updates are discussed below.

Pooling Data
Pooling data of similar components when updating the reliability performance is a technique advocated in Ref. D.1. This reference uses the following equations to calculate component unreliability.

$$UR_{Bc} = P_D + \lambda T_m \qquad \text{(NEI 99-02 Eq. 4)}$$

where:

P_D is the component failure on demand probability calculated based on data collected during the previous 12 quarters,

λ is the component failure rate (per hour) for failure to run calculated based on data collected during the previous 12 quarters, and

T_m is the risk-significant mission time for the component based on plant-specific PRA model assumptions.

$$P_D = \frac{(N_d + a)}{(a + b + D)} \qquad \text{(NEI 99-02 Eq. 5)}$$

where:

N_d is the total number of failures on demand during the previous 12 quarters,

D is the total number of demands during the previous 12,

and

a and b are parameters of the CNIP, derived from industry experience.

$$\lambda = \frac{(N_r + a)}{(T_r + b)} \qquad \text{(NEI 99-02 Eq. 6)}$$

where:

N_r is the total number of failures to run during the previous 12 quarters,

T_r is the total number of run hours during the previous 12 quarters.

As can be seen from Equations 5 and 6 above, if the number of demands or run hours is increased, then the impact of a single failure is reduced. For highly reliable components, the expectation is that the improvements in demand and run hours, due to data pooling, far outweigh the increase in failures due to the increased number of components that are monitored.

The application of data pooling is limited in that it requires the monitored components to be within a group of similar components and requires them to have significant demands and/or run hours to fully resolve the sensitive indicator issue. For a small pool of high-Birnbaum components, pooling data does not resolve the issue.

Merging Failure Modes for Turbine and Diesel-Driven Components

In addition to pooling data from similar components, data can be pooled by consolidating the various failure modes for a given component. This is only achievable if an appropriate unit for the reliability data can be determined. The failure mode merging technique works for turbine and diesel-driven components.

For turbine and diesel-driven components, extended run times (greater than 1 hour) are not typical, and the number of starts and the number of run hours are nearly equivalent. Since the historical failure rates are based on the failure to start and the failure to run, the combined failure rate is the sum of these failures. This assumes that the typical run duration is one hour.

The merging of the failure modes for turbine and diesel-driven components is one possible option for incorporation into the MSPI framework. The need for the frontstop would be reduced due to the reduced sensitivity of these components.

Data Update

The NEI MSPI methodology uses a posterior mean from updating a constrained noninformative prior (CNIP). See Ref. D.3 for a discussion of the CNIP. The following two alternative approaches have been investigated to address both invalid and insensitive indicators:

(1) Base the decision on percentiles of the posterior distribution rather than on the posterior mean.

Limited benefit was seen in the use of percentiles due to the nondiscriminatory nature of this approach. That is, both sensitive and insensitive indicators are impacted. Although sensitive indicator performance is improved, less sensitive indicators are made even less sensitive.

(2) Use a different prior, a mixture of two simple distributions corresponding to "normal" and "degraded" states, respectively.

Mixture priors are discussed in Ref. D.4. In exploratory work for this application, the use of a mixture prior showed good results for both sensitive and insensitive indicators. However, this method results in added complexity in both communicating the concept and in the implementation of the methodology. It is not immediately practical to implement the mixture prior.

For now, approaches that improve the data updating processes are not considered in the development of the frontstop.

D.1.3.1.2   High Fussell-Vesely Importance

As can be seen from Equation 3, component importance, normalized by dividing the importance by the unreliability, $^{FV}/_{UR}$, is a direct multiplier used for the determination of the change in risk due to a change in performance. Those components with high $^{FV}/_{UR}$ values are likely to be more sensitive to small changes in performance. The impact of the $^{FV}/_{UR}$ value is considered in the frontstop.

D.1.3.1.3   High CDF

CDF is a direct multiplier used for the determination of the change in risk due to a change in performance. Therefore, plants with a higher calculated CDF will have greater sensitivity to small changes. The influence of the calculated CDF is considered in the frontstop.

D.1.3.2   MSPI Unavailability Index (UAI)

The UAI uses a similar equation to that of the URI and can also be found in Ref. D.1. However, of interest to the frontstop design, is the development of the UAI's baseline unavailability. This baseline unavailability has two elements: planned and unplanned. Each element is derived from a different data source. The planned unavailability is the actual, plant-specific 3-year total planned unavailability for an in-scope train for the years 1999 through 2001. The baseline unplanned unavailability is the historical industry average for unplanned unavailability for the years 1999 through 2001. Basing planned unavailability on plant-specific practices is of interest since it directly relates current maintenance practices at the monitored plant to the baseline Plants that maintain these practices should not challenge the MSPI due to planned maintenance. However, changes in maintenance practices, especially due to the implementation of risk-informed allowed outage time (AOT) extensions, could impact the actual planned maintenance and may challenge the MSPI indicator. The impact of planned maintenance is further discussed in Section D.4.

D.2   Desired Frontstop Characteristics

A fundamental objective of the MSPIs is to monitor system performance so that declining performance is identified before it becomes unacceptable. Although the frontstop supports this objective, its focus is narrower. If the framework of the frontstop is appropriately constructed, then changes that are within a band of acceptable performance, including single failures, would not result in exceeding an action threshold. However, declining performance would be identified.

In order to achieve this fundamental objective, the following characteristics are considered critical for an effective frontstop:

- Address invalid indicators (thereby reducing false positives).
- Be compatible with, but not ignore, the Unavailability index contribution.
- Maintain sensitivity (without adversely impacting false negatives).

Each of these characteristics is discussed below.

D.2.1   Addresses Invalid Indicator

An important characteristic is that no single failure results in WHITE, that is makes $URI>1\times10^{-6}$. If invalid indicators are not eliminated, this would directly challenge the ability to work within the MSPI framework.

D.2.2 Compatible with Unavailability

An effective frontstop should be able to appropriately address a change in performance that results from a failure in light of any prior performance, whether at baseline or at some other state. This is especially true for how the frontstop relates to unavailability. Both the URI and UAI indices are impacted by failures. The URI contribution increases due to the updated failure rate while the UAI contribution increases due to the repair time required to return the failed component to service. In addition, both indices will reflect the system's performance prior to a failure. The frontstop must be able to address the interaction between unreliability and unavailability. It can not prevent the indicator from going WHITE if URI is near zero and UAI is greater than 1×10^{-6}.

D.2.3 Indicator Sensitivity is Maintained

The structure selected for the frontstop must maintain the MSPI's ability to identify degraded performance. The following criteria are considered to represent degraded performance:

- Two significant failures (each with a risk contribution greater than 5×10^{-7}) would very likely result in a WHITE indication.

- One significant failure with other less-significant failures could exceed the GREEN/WHITE threshold.

- One significant failure with a significant UAI contribution could exceed the GREEN/WHITE threshold.

D.3   Proposed MSPI Frontstop

The proposed MSPI frontstop places a cap on the URI contribution for the most significant failure in any 12-quarter reporting period at 5×10^{-7}. This risk cap ensures that two significant failures (i.e. failures contributing $>5 \times 10^{-7}$ to the URI) result in WHITE. It also ensures no invalid indicators, with some restrictions. Indicators that have a 5×10^{-7} failure contribution with $>5 \times 10^{-7}$ UAI will result in WHITE. Indicators that have a significant contribution from either URI or UAI, or both, prior to a significant failure may result in WHITE.

D.3.1 MSPI Frontstop URI Risk Cap of 5×10^{-7}

For the risk cap to be effective, its value needs to be less than 1×10^{-6} to prevent invalid indicators, and equal to or greater than 5×10^{-7} to maintain the MSPI sensitivity as discussed in Section D.2.3. Within this range, a risk cap of 5×10^{-7} is recommended for consistency with the current NRC position for a small quantitative impact for a single technical specification (TS) change.

RG 1.177, "An approach for Plant-Specific, Risk-Informed Decisionmaking: Technical Specifications" (Ref. D.5), includes an acceptance guideline for a small quantitative impact on plant risk attributable to a permanent TS change. It uses the metric of incremental conditional core damage probability (ICCDP).

> ICCDP = [(the conditional CDF with the subject equipment out of service) − (baseline CDF with nominal expected equipment unavailabilities)] x (duration of single AOT under consideration)

RG 1.177 states "An ICCDP of less than 5E-7 [5×10^{-7}] is considered small for a single TS AOT change." The ICCDP is very similar to the UAI calculation in that it evaluates the change in unavailability from the baseline to determine risk. Capping the MSPI risk associated with the most significant failure at 5×10^{-7} leaves a nominal 5×10^{-7} for the unavailability associated with the failure (assuming performance is at baseline). That is, the repair activities associated with this significant failure could result in an UAI contribution of 5×10^{-7} without exceeding the WHITE threshold. A higher risk cap, greater than 5×10^{-7}, would reduce the UAI margin for returning failed components to service. A lower margin could create a conflict between the MSPI and risk-informed AOT extensions. Such a conflict would occur if a licensee had received a risk-informed AOT extension allowing a one-time entry into a TS action statement based, in part, on the 5×10^{-7} guideline but had a more restricted MSPI limit. This leads to the question of whether the governing limit is the approved AOT extension or the MSPI. Conforming the proposed risk cap to RG1.177 ensures that when the licensee's performance is at baseline, the risk margins for risk-informed AOT extensions and the MSPI frontstop are consistent.

Note that RG 1.177 guidelines are intended for comparison with a full-scope (including internal events, external events, full power, low power and shutdown) assessment of the change in risk metric. Since the MSPIs only address internal events, the risk margin for unavailability is somewhat greater than it would be if a full-scope PRA was considered.

D.3.2 MSPI Frontstop UAI Unaffected

The UAI contribution is unaffected by the proposed frontstop. Its value is added to the frontstop-adjusted URI value to obtain the resulting MSPI value. Since the GREEN/WHITE threshold is 1×10^{-6} and the URI risk cap is 5×10^{-7}, there remains an approximate risk margin of 5×10^{-7}, potentially less if previous performance is worse than baseline and more if performance is better than baseline, to execute the repair activities. As stated in Section D.3.1, this is consistent with RG 1.177.

D.3.3 Indicator Sensitivity (White/Yellow Threshold)

The proposed frontstop only applies to the GREEN/WHITE threshold. If the calculated risk without the frontstop adjustment, exceeds the WHITE/YELLOW threshold of 1×10^{-5}, the adjustment is not applied. This approach maintains the basic criterion of the WHITE/YELLOW threshold.

D.4 Changes in Baseline UA

As discussed above, the MSPI frontstop does not limit the unavailability index for a given failure. Since the unavailability margin for repairs is directly impacted by the prior performance, it is important to ensure that planned unavailability is consistent with approved maintenance practices. This point is emphasized due to the changing nature of planned maintenance practices and the baselining of these practices to plant-specific data. A key planned unavailability change agent is risk-informed AOT extensions. Implementation of a risk-informed AOT extension could significantly change the plant-specific unavailability baseline. If the MSPI planned unavailability baseline were significantly different from the unavailability that results from an approved AOT extension, then the licensee would be forced to manage potentially conflicting expectations. It is therefore recommended that as part of the implementation of the proposed frontstop, adjustments to the baseline unavailability for planned unavailability be allowed for NRC-approved risk-informed changes.

D.5 Assessment of the Proposed Frontstop

This section evaluates the performance of the proposed frontstop with respect to the desired critical characteristic discussed in Section D.2.

Addresses Systems with Invalid Indicators

By limiting the risk of the most significant failure to 5×10^{-7}, an additional 5×10^{-7} remains for the sum of past performance plus repair unavailability. If past performance is at baseline, then a total of 5×10^{-7} is available for the risk associated with repair. Only if the repair unavailability is excessive or previous performance provides limited repair opportunity will a single failure result in exceeding the GREEN/WHITE threshold.

Compatible with Unavailability

Using a risk cap, as opposed to limiting the number of failures, gives the ability to directly interface the unreliability frontstop with unavailability. The limit on URI does not reduce the sensitivity of the MSPI to unavailability, it only reduces its sensitivity to a single risk significant failure. If an indicator has a near-zero URI, and UAI is greater than 1×10^{-6}, the indicator will be WHITE (or higher).

Indicator Sensitivity is Maintained

With a limit of 5×10^{-7} on delta URI on the most significant failure, a second significant failure of 5×10^{-7} or greater will result in at least a WHITE indicator. Other combinations of a 5×10^{-7} failure and lesser failures or unavailability greater than baseline would result in WHITE when the other failures and unavailability have a value that is greater than 5×10^{-7}.

The sensitivity of an indicator that has a value greater than 1×10^{-5}, the WHITE/YELLOW threshold, is not affected by the frontstop. That is, the proposed frontstop applies only to the GREEN/WHITE threshold. Moreover, once the GREEN/WHITE threshold has been exceeded, so that a WHITE performance has been declared, the risk cap would be waived and the numerical MSPI result would revert to what it would have been in the absence of the risk cap. In so doing, the frontstop only affects crossing of the GREEN/WHITE threshold, and has no effect on higher thresholds.

D.6 Examples Using the Proposed Frontstop

The following cases are sample applications of the proposed frontstop.

D.6.1 Case 1

Scenario

A plant experiences a start failure of an Auxiliary Feedwater motor-driven pump. Prior to the failure, the UAI = 1×10^{-7}. The delta URI associated with the start failure is 4×10^{-6}. No other failures have occurred during this reporting period yielding an URI baseline of zero (this is a simplification since baseline could be below zero). The UAI contribution resulting from the repair unavailability is 2×10^{-7}.

MSPI Calculation

Without the frontstop, the MSPI would be an invalid WHITE (sometimes denoted "GRAY") with a resulting value of 4.3×10^{-3}. With the frontstop, the failure is limited to delta URI of 5×10^{-7} (the risk cap) that is added to the previous UAI of 1×10^{-7} and the repair contribution to UAI of 2×10^{-7}. This results in a total MSPI of 8×10^{-7} (GREEN).

D.6.2 Case 2

Scenario

A plant experiences a start failure of an auxiliary feedwater motor-driven pump. Two previous failures have occurred during this reporting period. One was on this same pump and a second was on a motor-operated valve failing to open on demand. The delta URI associated with the start failures is 4×10^{-6} each. The delta URI of the motor-operated valve (MOV) failure is 1×10^{-7}. The previous UAI, which includes the first two failures, is 2×10^{-7}. The delta UAI resulting from the repair of the current failure is 2×10^{-7}.

MSPI Calculation

Without the frontstop, the MSPI would be the sum of the two AFW pump failures, 8×10^{-6}, plus the MOV failure for a total URI of 8.1×10^{-6}. UAI would sum to 4×10^{-7} and the total MSPI would be 8.5×10^{-6} (WHITE). With the frontstop, the most significant failure would be reduced to a delta URI of 5×10^{-7}. Since there are two failures of equal risk, one of these two would be reduced by the risk cap. This would result in an URI of $5 \times 10^{-7} + 4 \times 10^{-6} + 1 \times 10^{-7} = 4.6 \times 10^{-6}$. This is added to the UAI, which is unchanged at 4×10^{-7}, for a total possible MSPI of 5×10^{-6} (WHITE). In this case, because the performance would have been declared WHITE regardless of the application of the risk cap, the frontstop is waived and the MSPI would revert to 8.5×10^{-6} as if there were no risk cap. Thus, the margin to the WHITE/YELLOW threshold is not affected by the frontstop.

D.6.3 Case 3

Scenario

A plant experiences a start failure of an Auxiliary Feedwater motor-driven pump. Two previous start failures have occurred during this 12-quarter reporting period on the same pump. The delta URI associated with the start failures is 4×10^{-6} each. The previous UAI, which includes the first two failures, is 2×10^{-7}. The delta UAI resulting from the repair of the current failure is 2×10^{-7}.

MSPI Calculation

Without the frontstop, the sum of the three AFW pump failures results in delta URI of 1.2×10^{-5}. The unavailability contribution (UAI) is 4×10^{-7}. This yields a total MSPI of 1.24×10^{-5} (YELLOW). Since the frontstop only applies to the GREEN/WHITE threshold, the resulting MSPI would remain at 1.24×10^{-5} (YELLOW) as first calculated without the frontstop.

D.7 References

D.1 Nuclear Energy Institute (NEI). NEI 99-02 (Draft Report), "Regulatory Assessment Performance Indicator Guideline," Section 2.2 ("Mitigating Systems Performance Index") and Appendix F ("Methodologies for Computing the Unavailability Index, the Unreliability Index, and Determining Performance Index Validity"). NEI: Washington, DC. 2002.

D.2 NRC Interoffice Memorandum from Scott F. Newberry (RES/DRAA) to John A. Zwolinski (NRR), "Request for Review of Mitigating Systems Performance Indices White Paper, ADAMS Accession Numbers ML031350208 and ML031360121, May 12, 2003.

D.3 C.L. Atwood, "Constrained Noninformative Priors in Risk Assessment," *Reliability Engineering and System Safety*, Vol. 53, No, 1, pp 37–46, 1996.

D.4 C.L. Atwood and R.W. Youngblood, "Application of Mixture Priors to Assessment of Performance," *Probabilistic Safety Assessment and Management (PSAM 7-ESREL '04)*, pp. 444-449, edited by Cornelia Spitzer, Ulrich Schmocker, Vinh Dang (Springer), 2004.

D.5 U.S. Nuclear Regulatory Commission, Regulatory Guide 1.177, "An Approach for Plant-Specific, Risk-Informed Decisionmaking: Technical Specifications," Washington, DC, August 1998.

APPENDIX E. TECHNICAL BASIS FOR THE BACKSTOP TO ADDRESS INSENSITIVE INDICATORS

Appendix E
Technical Basis for the Backstop to
Address Insensitive Indicators

E.1   Introduction

Although the systems selected for monitoring are relatively risk-significant at most plants, the Birnbaum measures (Bs) for specific system *trains* may be relatively small numbers at some plants. This is attributable, in part, to the system selection process — an indicator defined for systems that are important at many plants, but not at all plants, may be *insensitive* at some plants. A low value of train B can also easily arise in highly redundant systems; failure of *individual* trains in a highly redundant system may not yield a high conditional CDF, even if failure of the entire system would do so. In such a case, the number of failures needed to produce a change in the MSPI greater than 1×10^{-6} is large. This makes it possible for many failures to occur in a system having apparent regulatory significance, with the performance index still falling short of the WHITE performance band threshold.

This is undesirable from both technical and outside stakeholders' points of view. From an outsider perspective, an indicator scheme appears deficient if large numbers of failures do not warrant a "WHITE" response. Moreover, absent a comprehensive model relating licensee performance to different kinds of indications, it is difficult to conclude on purely technical grounds that such performance excursions are risk-insignificant, even if they arise in low-B trains. Examples of this are the following. First, the occurrence of an unexpectedly large number of failures implies a performance issue that could well be cross-cutting (i.e., affecting other systems), and have a net effect on ΔCDF that is somehow not captured in the current calculations. Second, a performance issue causing a large number of failures could easily alter the effective common-cause failure (CCF) parameters. The current approach of NEI 99-02 Appendix F (Ref. E.1) does not explicitly update the effective CCF parameters, so the risk significance of a performance issue affecting the CCF parameters can be understated by the current calculational approach.

Therefore, it is desirable to supplement the 1×10^{-6} threshold criterion for entry into "WHITE" with another criterion. This criterion will be based on the statistical significance of the observed number of failures, relative to prior expectations. When a number of failures is observed larger than or equal to a specified "backstop" value, a WHITE will be declared, independently of the calculated change in the MSPI.

When evaluating a backstop, it must be recognized that baseline conditions include both normal performance and degraded performance, with normal performance occurring in the vast majority of the cases. A "positive" indicator consists of a failure count at or above the backstop. It is a "false positive" if the underlying performance is normal (with the many failures having been just the result of coincidence), and a "true positive" if the underlying performance is degraded. The backstop threshold will be formulated to have the following properties:

- The false positive rate will be low. This criterion can be formulated to say that the conditional probability of declaring "WHITE," given normal performance, will be very low (actual cutoff probability determined below). This is the classical notion of hypothesis testing, based on the consistency of the data with "normal" performance.

- Of all the positives that occur under baseline conditions, only very few are *false* positives. This criterion involves both the probability of false positives and the probability of true positives, under the *a priori* baseline conditions. Thus, a "WHITE" will be declared only when the number of observed failures leaves little room for doubt regarding the existence of a performance issue.

These two objectives can be satisfied by adjusting the backstop threshold to correspond to the smallest possible number of failures, consistent with achieving the desired low false positive rate. This is discussed below.

Because the "backstop" is intended to address failures that are in some sense repetitive, comparison with the intent of the Maintenance Rule is natural. There is one key similarity between the Maintenance Rule and the MSPI with backstop: an unexpectedly high number of failures triggers corrective action. The intent of the present MSPI backstop development is to formulate backstops that envelop licensee goals under Maintenance Rule implementation. That is, licensees will ordinarily trip their own Maintenance Rule goals before they trip the backstop. At some plants, it may be possible for a peculiar sequence of failures to trip the MSPI backstop first. However, since the MSPI backstop will be designed with a low false positive rate, this is not necessarily undesirable; it may signal a performance issue that is real enough, despite having gotten past the Maintenance Rule criterion.

The effect of the "frontstop" and "backstop" on the decision rule for declaring "WHITE" is illustrated in Figure E.1 below.

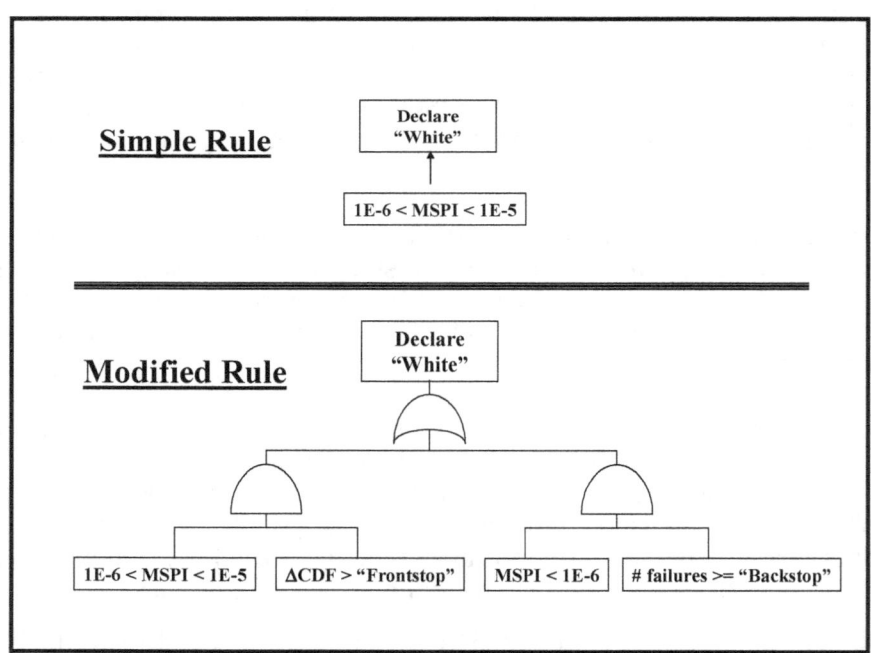

Figure E.1 Decision Rule for Declaring WHITE with Backstop

E.2 Parameter Distributions

Conceptually, the "backstop" is a limit on the total number of failures, of all failure modes and of all components of one type in one system of a single nuclear power plant unit. Each system and type of component corresponds to a single backstop, with all failure modes combined If the number of failures seen in the 3-year performance period is equal to the backstop number or more, the system/component has reached or exceeded the backstop and is denoted "WHITE." The details of this definition are given below.

The criteria defining the backstop are based on probabilities. These probabilities are predicated on a realistic belief about the possible values of the failure parameters, recognizing that a single parameter will be somewhat different at different plants and in different time periods. Each parameter was assigned a distribution, reflecting belief about the values that the parameter could actually take at various plants in various 3-year periods. The distribution was developed as follows.

The data were collected for the following systems/components/failure modes:

Air-operated valves	Failure to operate
Circuit breakers	Failure to open or close
EDG circuit breakers	Failure to close
Emergency diesel generators	Failure to load and run
Emergency diesel generators	Failure to run
Emergency diesel generators	Failure to start
Motor-driven pumps, norm. running	Failure to run
Motor-driven pumps, norm. running	Failure to start
Motor-driven pumps, standby	Failure to run
Motor-driven pumps, standby	Failure to start
Motor-operated valves	Failure to operate
Turbine-driven pumps, AFW	Failure to start
Turbine-driven pumps, all	Failure to run
Turbine-driven pumps, HPCI/RCIC	Failure to start

Diesel-driven pumps were not considered here, because they are present at very few plants. For each system/component/failure mode, the data were collected separately in two 3-year periods, 1997 – 1999 and 2000 – 2002. The reason for using 3-year periods is that the MSPI pilot program will look at 3-year windows of data. Therefore, it is most relevant to use comparable windows of data in the analysis here.

For each system/component/failure-mode/data-period, the empirical Bayes (EB) distribution was found, modeling between-plant variation in either p (for failure to start, failure to load and run, or failure to open or close) or λ (for failure to run). The plant-specific means were tabulated. Each plant-specific mean is a "best estimate" of the parameter at the plant during the 3-year period. In particular, it is better than the maximum likelihood estimate (MLE) using the plant-specific data, because plants with few demands or few exposure hours do not have as great a volatility in their EB posterior means as in the MLEs.

For each system/component/failure-mode/data-period, the empirical Bayes means were rescaled by dividing them by the industry mean. This put all the parameters on the same scale, with mean 1. For two system/component/failure-mode/data-periods, the empirical Bayes distribution was degenerate, showing no between-plant variability. For these two cases, every plant-specific parameter was assigned a rescaled value of 1.

Plots were examined, and no correlations were evident. That is, for any system/component/failure-mode, a plant that was high in one 3-year period did not show a tendency to be high in the other 3-year period. Also no plant seemed to be consistently high for more than one system/component/failure-mode. Therefore, the results for different system/component/failure-mode/data-periods were treated as independent of each other.

The rescaled plant-specific means were pooled into a single data set, with 2388 values. The smallest value was 0.016 and the largest value was 24.05. At this point, the distinction between ps, having beta distributions, and λs, having gamma distributions, was ignored. This is not unreasonable, because a beta distribution with small mean is approximately a gamma distribution.

The values were ordered from smallest to largest, and the empirical cumulative distribution was plotted. This is shown in Figure E.2.

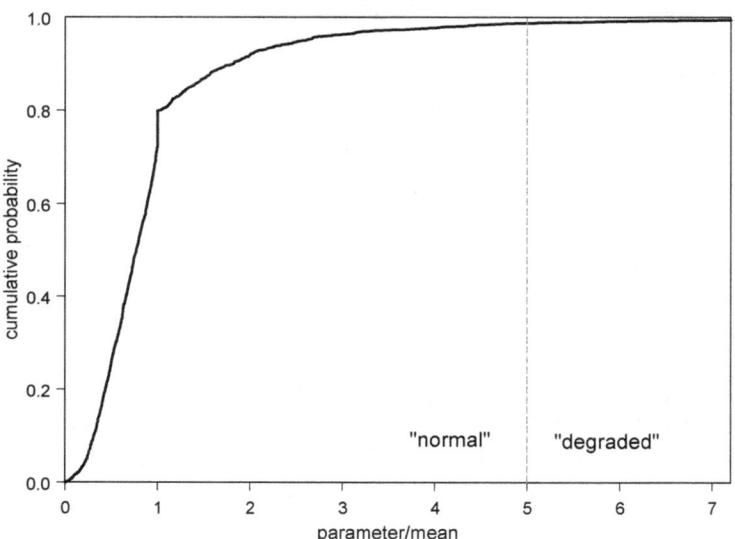

Figure E.2 Empirical Distribution of Rescaled Plant-Specific Parameters

A little bump can be seen corresponding to the cases when the EB distribution was degenerate and the rescaled plant-specific means were set to exactly 1. That bump is an artifact of the EB methodology.

Each parameter was assigned a generic probability distribution based on this distribution. That is, each parameter has an industry mean, obtained from industry experience in 1997 – 2002. For any particular parameter, the distribution in Figure E.2 was rescaled so that its mean was the industry mean of the parameter. The resulting distribution was used for the parameter.

E.3   Positives, False and True

Parameter values less than 5 times the industry mean were considered "normal." Values that are more than 5 times the industry mean were considered "degraded." This dichotomy into good and degraded is shown in Figure E.2. There were 30 "degraded" points in the distribution of 2388 points. Therefore, without seeing data, one can set a priori Pr(parameter is degraded) = 30/2388 = 0.0126 — any parameter is probably normal almost all of the time at almost all of the plants.

Suppose that some candidate value has been nominated as a backstop, for some system and component type. If the observed failure count equals the backstop or more, call this a "positive." For example with a pump, the total count of failures to start and failure to run would be compared with the backstop limit. A "false positive" is a case when all the corresponding parameters (p_{FTS} and λ_{FTR} in the example) are normal yet the count equals the backstop or higher. A "true positive" is a case when at least one of the parameters is degraded and the count equals the backstop or more.

Pr(false positive) therefore is the probability that two things occur: all the parameters are normal and the data count is as high as the backstop or higher.

Pr(false positive) = Pr(parameters normal)×Pr(backstop exceeded | parameters normal).

The second factor on the right-hand side is the conditional probability, given that the parameters are normal. In classical hypothesis testing, only this conditional probability is considered. However, the value of Pr(parameters normal) is treated as known in the present work — for example, it is $(1 - 0.0126)^2$ for two parameters each having distributions based on Figure E.2. Therefore, any criterion based on the unconditional Pr(false positive) is equivalent to a criterion in terms of the conditional probability.

E.4   Precise Definition of Backstop

The backstop was chosen to be the smallest number such that:

(1) Pr(false positive) ≤ = 0.01
(2) Pr(false positive)/Pr(positive) ≡ fraction of positives that are false ≤ 5%.

Thus, the backstop is defined to ensure that false positives are very rare, and if a positive occurs it is very probably a true positive. By the last paragraph of the previous section, the first condition can be re-expressed in terms of Pr(backstop exceeded | normal parameters), the conditional probability that is used in hypothesis testing. The second condition was the more difficult condition to fulfill, and governed the value of the backstop limit in every case.

The calculations depend on the assumed distribution for the parameters and on the number of demands or the running time for the components in the particular system in a 3-year period at the plant. If two otherwise-identical plants have different demands counts and run-times, the one with more demands and run hours may have a higher backstop limit.

E.5 Calculation Method

Because the underlying probability distribution of the parameter values (Figure E.2) was discrete, and the number of failures is a discrete distribution (Poisson or binomial) depending on the parameter value and on the total demand count or run time, all the calculations could be performed in a spreadsheet. In the pump example, the equations are as follows:

First,
Pr(x failures to start and p normal) = Σ_i Pr(x failures to start | p_i) Pr(p_i),
where the probability distribution of X is binomial, the distribution of p is based on Figure E.2, and the sum is over all i in the "normal" part of Figure E.2.

Similarly,
Pr(y failures to run and λ normal) = Σ_i Pr(y failures to run | λ_j) Pr(λ_j),
where the conditional distribution of Y is Poisson, and the distribution of λ is based on Figure E.2.

The probability of z failures when both parameters are normal is given as follows:
Pr(z failures and both parameters normal)

$$= \sum_{x=0}^{z} \text{Pr}(x \text{ failures to start and } p \text{ normal}) \, \text{Pr}(z - x \text{ failures to run and } \lambda \text{ normal}).$$

Finally, for a candidate backstop b, the probability of a false positive is

$$\text{Pr(false positive)} = 1 - \sum_{z=0}^{b-1} \text{Pr}(z \text{ failures and both parameters normal}).$$

The calculations are all based on equations such as these. For each candidate value of the backstop, the probability of a false positive and the fraction of positives that are false were calculated. The value selected as the backstop was the smallest candidate backstop satisfying constraints 1 and 2 above.

E.6 Backstop Values

The backstops were first calculated on a system basis for the major components at all 20 pilot plants. Mean values and standard deviations were next generated based on similar component types. Table E.1 below gives the backstops if generic values were to be used. The standard deviations are shown for information only, and provide a measure of how much plant-to-plant variability there is.

Table E.1 Generic Backstops

Component	mean	st dev
AOV	5	0.9
DDP	13	4.5
EDG	9	1.7
MDP	7	1.4
MDP Stby	6	2.7
MOV	5	1.1
TDP	6	1.0

The expected number of failures based on the number of demands and run hours for each component type within a system for all pilot plants can be derived. The expected failure count is

$$pD + \lambda t$$

where p = Pr(failure on demand), D = number of demands, λ = rate of failure to run, and t = number of run hours. When generic failure rates were used, a strong correlation was observed between backstop and expected number of failures, for each component type.

One could also plot the backstop versus expected failure count for *all* component types on a common graph, and still observe a strong correlation. Figure E.3 below shows this correlation. In practice, calculated values using the linear regression expression would be rounded up or down to the nearest integer. For example, assume a particular plant had two similar standby motor-driven auxiliary feedwater pumps. Use of the total number of start and run hours over a 3-year period, in combination with the revised generic mean failure rates (Table C.2), would allow the derivation of the expected number of failures for the two pumps. The equation in Figure E.3 would yield a number y which, after being rounded to the nearest integer, would be the backstop for the two pumps. This process could be applied to all similar components within a system, for all systems in the MSPI. The advantage of Figure E.3 over Table E.1 is that the variable backstop allows for the variation in design configuration (number of components), testing frequency, and operation.

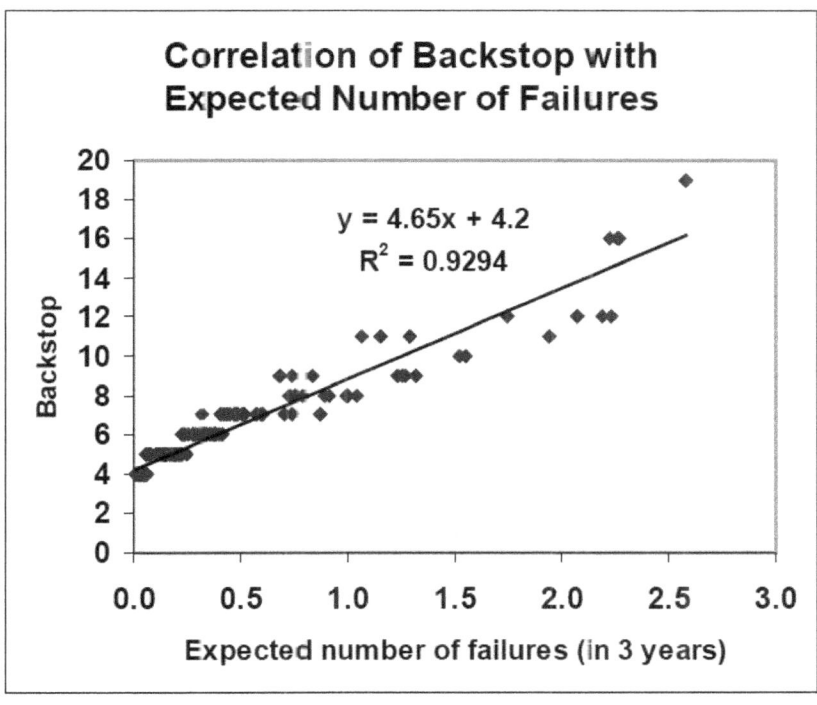

Figure E.3 Variable Backstop

E.7 A Short-Term Backstop

The effect of using a "short-term backstop" was investigated. This might be implemented if it were desirable to catch incipient problems quickly, without waiting up to 3 years. The definition is the same as for the backstop given above, except the time period for collecting data is only six months. Thus, if too many failures of any system have occurred in the previous two quarters, the system is declared to have exceeded the short-term backstop, and appropriate action or investigation could be initiated.

A least-squares fit was constructed, just as above. However, the fitted line did not follow the data points as closely as in Figure E.3, primarily because the expected number of failures had a very short range, from 0.0 to only about 0.4. The value of the short-term backstop was 4 or 5 for nearly all systems at nearly all plants. This is illustrated in Figure E.4.

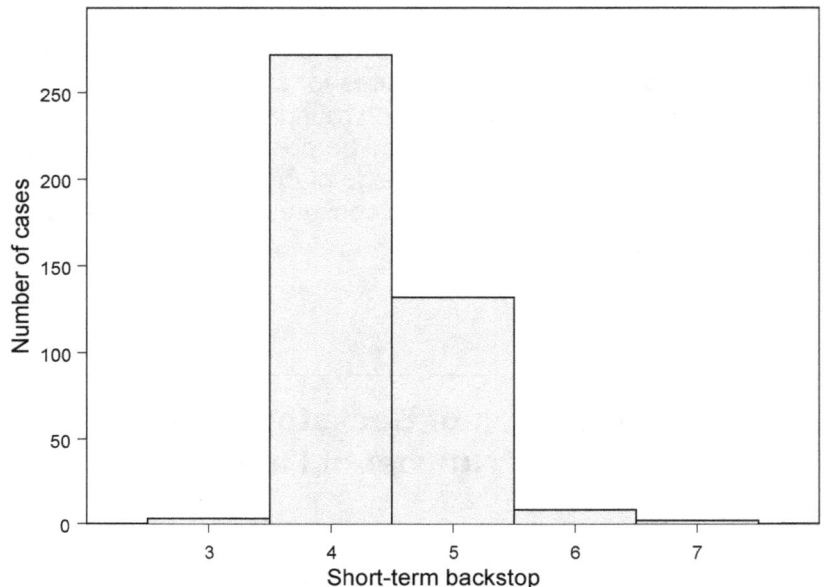

Figure E.4 Values of Short-Term Backstop

Therefore, the simplest short-term backstop is 4: if 4 or more failures occur in two consecutive quarters in one system, declare the backstop to have been exceeded. This could be refined by noting that the exact short-term backstop was typically 5 for MDPs and typically 4 for other systems, and that certain identified plants were outliers. However, the "one size fits all" short-term backstop would be 4 or more failures in 2 quarters.

No recommendation is made regarding implementation of the short-term backstop, and it remains an option for future consideration.

E.8 References

E.1   Nuclear Energy Institute (NEI). NEI 99-02 (Draft Report), "Regulatory Assessment Performance Indicator Guideline," Section 2.2 ("Mitigating Systems Performance Index") and Appendix F ("Methodologies for Computing the Unavailability Index, the Unreliability Index, and Determining Performance Index Validity"). NEI: Washington, DC. 2002.

APPENDIX F. TECHNICAL BASIS FOR TREATMENT OF COMMON-CAUSE FAILURE CONTRIBUTION TO FUSSELL-VESELY IMPORTANCE

Appendix F
Technical Basis for Treatment of Common-Cause Failure Contribution to Fussell-Vesely Importance

F.1 Introduction

This appendix provides a methodology for adjusting the Mitigating Systems Performance Index (MSPI) Unreliability Index terms proposed by NEI (Ref. F.1) to address the common-cause failure (CCF) contribution to these indices. Specifically, it addresses the impact of a change in the independent failure probability on the CCF probability. It does not address the impact of changes in the CCF parameters.

The current NEI proposal is to account only for "independent" failures in the MSPI. The NEI-proposed approach would not account for the contribution to common cause attributable to a change in total reliability.

The present approach to address the CCF contribution provides a first order mathematical approximation. It requires one input beyond those already required by the MSPI, namely, the Fussell-Vesely (FV) importance value of the CCF event associated with each in-scope common-cause group.

Conceptually, the use of the FV as a factor to adjust the MSPI for common cause appears reasonable. This factor directly addresses the importance of the common-cause contribution. However, two other factors need to be considered. These are the degree of redundancy and the degree of common-cause coupling. Both of these issues are also addressed by this approach through the $FV_{(CommonCause)}/UR_{(Independent)}$ ratio (UR refers to component unreliability) inherent in the MSPI equation. This is described in detail later in this appendix.

Note that this common-cause adjustment only addresses the impact of changing the independent failure rate on common-cause failure rate in PRA models, and hence on the MSPI. It does not attempt to address the conditional risk associated with a multiple-failure event attributable to common cause. The current program position is that while total failure counts would go into the MSPI, multiple-failure events *per se* would be addressed through the inspection process.

F.2 Methodology

This section develops the methodology for applying common cause to the MSPI.

F.2.1 MSPI System Unreliability Index (*URI*)

Equation 3 of Reference F.1 defines the system Unreliability Index (*URI*) and is reproduced below. This equation is modified later in this appendix to reflect the impact of common cause on CDF.

$$URI = CDF_p \sum_{j=1}^{m} \left[\frac{FV_{URcj}}{UR_{pcj}} \right]_{max} (UR_{Bcj} - UR_{BLcj}),$$

(NEI 99-02 Eq. 3)

where the summation is over the number of active components (*m*) in the system, and:

CDF_p is the plant-specific internal events, at power, core damage frequency,

FV_{URc} is the component-specific Fussell-Vesely value for unreliability,

UR_{Pc} is the plant-specific PRA value of component unreliability,

UR_{Bc} is the Bayesian corrected component unreliability for the previous 12 quarters, and

UR_{BLc} is the historical industry baseline calculated from unreliability mean values for each monitored component in the system.

F.2.2 Common-Cause Models

In order to clarify the relationship between independent failure probability and common-cause probability, a brief discussion of common-cause models is provided.

The Beta Factor, Multiple Greek Letter, and Alpha Factor models are typically used to quantify common-cause failure probabilities. These are "parameter" models: they use parameters based on ratios of common-cause failures to total failures from one source of data and a total failure probability from another source (Ref. F.2). It is this model structure that results in the change in common-cause failure probability for a given change in total failure probability. Although it is recognized that both the common-cause ratios and the total failure probability change with new data, the current proposal for MSPI does not attempt to quantify changes in the CCF model parameters. In effect, the common-cause ratios are considered constant over the limited range for which the independent failure rate changes are evaluated.

Within the Alpha Factor Model, the following relationship exists between the total and independent failure probabilities:

$$UR_{Independent} = \alpha_1 \times UR_{Total} \qquad \text{(Eq. F.1)}$$

where

$\alpha_k =$ fraction of the total frequency of failure events that occur in the system and involve the failure of *k* components due to a common cause.

There is a similar relation between the total and common-cause failure probabilities. This relationship can be simple, like that for a two component common-cause group:

$$UR_{CommonCause} = \alpha_2 \times UR_{Total} . \qquad \text{(Eq. F.2)}$$

Or it can be significantly more complicated, like that for a four component common-cause group where one of four must operate:

$$UR_{CommonCause} = \left(\tfrac{1}{3} \alpha_2^{\,2} UR_{Total} + \tfrac{4}{3} \alpha_1 \alpha_3 UR_{Total} + 2\alpha_1 \alpha_2 UR_{Total} + \alpha_4 \right) \times UR_{Total} \qquad \text{(Eq. F.3)}$$

The more complex relationships like that above are the reason that a first order approximation is needed for MSPI purposes. Without this approximation, the equation fragment that is multiplied with UR_{Total} would be also dependent on UR_{Total}. Since UR_{Total} is typically much smaller than the common-cause coupling factor, α_4 in this case, simplifying this equation to a form that is first-order in UR introduces minimal error.

$$UR_{CommonCause} \approx \alpha_4 \times UR_{Total} \qquad\qquad \text{(Eq. F.4)}$$

This can be generically represented by the following equation:

$$UR_{CommonCause} \approx \alpha_{CCF} \times UR_{Total} \qquad\qquad \text{(Eq. F.5)}$$

The generic form of this equation reflects the reparametrization form of the typical common-cause model and clearly shows the dependence of common cause on the total failure probability.

F.2.3 Component Unreliability (*UR*)

The NEI document defines UR_{pc} as the plant-specific PRA value of component unreliability. Typically, the failure rate used in the PRAs includes both independent and common-cause failures. That is, failures are not evaluated and screened due to their association with a common-cause event. The assumption is that UR_{pc} represents the total failure rate as opposed to the independent failure rate. In practice, the difference between these two values is small since most failures are independent. This clarification is necessary in order to establish an effective framework that addresses both independent and common-cause impacts.

F.2.4 URI Equation with Common Cause

Given the assumption that there is a change in both independent and common-cause failure probabilities as the result of a change in the total failure probability, the *URI* equation can be rewritten as follows:

$$URI_{Total} = URI_{Independent} + URI_{CommonCause} \qquad\qquad \text{(Eq. F.6)}$$

Since the NEI *URI* equation assumes that a change in component reliability has only an independent impact, one can equate the $URI_{Independent}$ with the current NEI *URI* equation. As noted above, this is slightly conservative in that the change on component unreliability includes both independent and common-cause failures. However, the more significant issue is the FV value used in the equation and this reflects only the independent impact. The common-cause impact is addressed by the second term in the above equation, $URI_{CommonCause}$.

Using Equation F.6 above, and NEI Equation 3, the following *URI* equation can be developed.

$$URI_{CCGroup} = CDF_p \sum_{j=1}^{m} \left[\frac{FV_{URcj}}{UR_{pcj}} \right]_{max} (UR_{Bcj} - UR_{BLcj})$$

$$+ CDF_p \left[\frac{FV_{URc(CommonCause)}}{UR_{pc(CommonCause)}} \right]_{max} (UR_{bc(CommonCause)} - UR_{BLc(CommonCause)})$$

(Eq. F.7)

For simplicity, this equation represents the components associated with a single common-cause group. This avoids having additional nomenclature to associate the common-cause group with its independent failures. Substituting $\alpha_{CCF} UR_{Total}$ for $UR_{CommonCause}$ from Equation F.5 yields the following:

$$URI_{CCGroup} = CDF_p \sum_{j=1}^{m} \left[\frac{FV_{URcj}}{UR_{pcj}} \right]_{max} (UR_{Bcj} - UR_{BLcj})$$

$$+ CDF_p \left[\frac{FV_{URc(CommonCause)}}{UR_{pc(CommonCause)}} \right]_{max} (\alpha_{CCF} UR_{Bcj} - \alpha_{CCF} UR_{BLcj})$$

(Eq. F.8)

Note that since the components within a common-cause group are *a priori* similar, their failure data are pooled in accordance with NEI (and NRC) guidance. This results in the same *ΔUR* being used for each component. This also results in the same *ΔUR* being used for the common-cause contribution, although this change is modified by the α_{CCF} factor. It can also be seen that the magnitude of the change in common cause is significantly less due to the presence of the α_{CCF} factor. This factor carries the knowledge of the degree of common-cause coupling and the degree of redundancy. The overall change in common-cause unreliability increases with increased coupling and decreases with increased redundancy. Therefore, all three common-cause characteristics are addressed: importance by the use of *FV*, as well as coupling and redundancy by use of the α_{CCF} factor.

Equation F.8 can be rewritten as follows:

$$URI_{CCGroup} = CDF_p \left(\sum_{j=1}^{m} \left[\frac{FV_{URcj}}{UR_{pcj}} \right]_{max} + \left[\frac{\alpha_{CCF} FV_{URc(CommonCause)}}{UR_{pc(CommonCause)}} \right]_{max} \right) (UR_{Bcj} - UR_{BLcj})$$

(Eq. F.9)

Using a modified version of Equation F.5, $UR_{(CommonCause)}$ can be substituted out of the equation.

$$UR_{(Independent)} = \frac{1}{\alpha_{CCF}} \times UR_{(CommonCause)}$$

(Eq. F.10)

This results in a new *URI* equation that only requires addition of the common-cause *FV* value as shown below.

$$URI_{CCGroup} = CDF_p \left(\frac{\sum_{j=1}^{m} FV_{URcj} + FV_{URc(CommonCause)}}{UR_{pc}} \right)_{max} (UR_{Bcj} - UR_{BLcj}) \qquad \text{(Eq F.11)}$$

This equation represents the *URI* for a given common-cause group. It can be generically represented by the following:

$$URI_{Total} = CDF_p \sum_{i=1}^{n} \left(\frac{\sum_{j=1}^{r} FV_{URcj} - FV_{URc(CommonCause)i}}{UR_{pc}} \right)_{max} (UR_{Bcj} - UR_{BLcj}) \qquad \text{(Eq F.12)}$$

$$+ CDF_p \sum_{j=1}^{m} \left[\frac{FV_{URcj}}{UR_{pcj}} \right]_{max} (UR_{Bcj} - UR_{BLcj})$$

where:

I indexes common-cause groups,

n represents the number of in-scope common-cause groups,

m represents the number of independent components (not associated with a common-cause group),

j indexes components within a common-cause group,

r represents the number of components within a given common-cause group.

The two parts of this equation are necessary to address both components associated with a common-cause group and components that are unique and independent of any common-cause group.

F-5

F.2.5 Example

As an example, consider a common-cause group of two emergency diesel generators. This would result in the following equation:

$$URI_{TOTAL} = CDF_p \left(\frac{FV_{UR(EDG1)} + FV_{UR(EDG2)} + FV_{UR(EDGCommonCause)}}{UR_{pc}} \right) \times \left(UR_{Bc(EDG)} - UR_{BLc(EDG)} \right)$$

Substituting values from the Palo Verde enhanced SPAR model this equation becomes:

$$URI_{TOTAL} = 1.2E - 5 \times \left(\frac{1.7E - 02 + 2.3E - 02 + 1.1E - 02}{1.5E - 02} \right) \times \left(UR_{Bc(EDG)} - UR_{BLc(EDG)} \right)$$

$$URI_{TOTAL} = 4.1E - 05 \times \left(UR_{Bc(EDG)} - UR_{BLc(EDG)} \right)$$

This can be compared to the *URI* equation that only considers independent failures by removing the common-cause *FV*.

$$URI_{Independent} = 3.2E - 05 \times \left(UR_{Bc(EDG)} - UR_{BLc(EDG)} \right)$$

In this case, there is a 30% increase in the *ΔUR* multiplier. This increase will vary depending on the common-cause importance, the degree of coupling and the degree of redundancy. Additional examples are shown below.

Table F.1 Examples of the Effect of Common Cause

Indicator	Plant	Component	Redundancy	UR	Sum of Component FV	CCF FV	Increase
RHR	Millstone 2	MDP Cntmt Spray	2	5.46E-03 (FTS)	1.21E-04	5.05E-06	5%
SWS	S. Texas	Pumps	3	1.32E-04 (FTR)	3.1E-03	1.9E-04	6%
EAC	Millstone 2	EDGs	2	8.02E-03 (FTS)	7.09E-03	7.58E-04	11%
EAC	S. Texas	EDG	2	8.26E-03 (FTLR)	4.09E-02	1.57E-01	17%
EAC	S. Texas	EDGs	3	3.17E-02 (FTR)	1.57E-01	2.6E-02	17%
EAC	Palo Verde (Enhanced SPAR)	EDGs	2	1.5E-02 (FTR)	4.0E-02	1.1E-02	30%
HPI	Millstone 2	MDP	3	3.36E-03 (FTS)	5.61E-02	2.07E-02	37%
SWS	Hope Creek	MDPs	4	5.47E-04 (FTR)	4.33E-03	7.53E-03	63%
EAC	Hope Creek	EDGs	4	6.83E-03 (FTR)	6.13E-03	3.94E-02	146%
EAC	Limerick	EDGs	4	1.19E-02 (FTR)	2.57E-02	2.40E-01	930%

As stated in the introduction, this equation does not capture the change in reliability resulting from a common-cause induced, multiple failure event. Such an equation would require that $FV_{UR(EDGCommonCause)}$ be divided by the UR for the common-cause basic event, a much smaller number that is reduced by the α_{CCF} factor. In addition, the change in UR would need to reflect the change in the coupling factor as well as the change in the independent failure likelihood.

F.2.6 Truncation

The truncation limit used during model quantification could have a significant impact on this approach for adjusting the MSPI equation to address common cause. Given the low common-cause failure probabilities when compared with the independent failure probabilities, there is a greater risk that a significant number of common-cause cutsets or sequences will be truncated at a given truncation level. This results in lower common-cause FV values and, therefore, an underestimation of the common-cause impact. This needs to be considered in determining the importance of the CCF basic event.

F.3 Process for Evaluating CCF Contribution to MSPI

The process for evaluating the CCF contribution to the MSPI is described below. This process addresses the various means by which common cause is treated in PRA models. The premise of this process is that the observed failure data relate to total failure probability. When total failure probability increases, so too does CCF probability, as implied by the parametric models commonly used. The risk significance of declining reliability performance is therefore affected by the risk significance of CCF. A flowchart of the overall process is shown in Figure F.1.

Step 1: For each component, determine whether it is within one or more CCF groups.

Common cause should be considered for components of similar design, operation, maintenance practices or environment. In accordance with the NEI guidance, demands and failures for similar components within each system are summed. Components that have been grouped for this purpose should be considered for common cause.

Step 2: For each common-cause group, determine the failure mode used for the maximum FV/UR.

The MSPI process only uses the failure mode with the maximum FV/UR for components within scope. The CCF associated with the failure mode is used to represent the impact of common cause on the MSPI.

Steps 3 and 3.1: Identify the associated CCF events within the PRA

For the identified failure mode, the associated CCF events that are modeled within the site-specific PRA should be identified. If there are no CCF events, then the appropriate event(s) should be added to the PRA. Alternatively, the lack of common-cause modeling should be justified and documented.

Step 4: Determine the modeling approach

PRA practitioners use a variety of techniques to apply CCF to fault-tree and event-tree models. The capability or limitations of the PRA software used for the models sometimes drive these techniques. Several different modeling approaches are discussed below. The overriding principle of all these approaches is to identify the total risk contribution from both independent and common-cause failures for each in-scope MSPI component.

Step 4.1: Single Event

Often, CCF is modeled as a single event that addresses all the combinations of the failures with exception of independent failures failing with other independent failures. These independent failures are modeled as separate basic events. For example, consider a system with three redundant components. In a "single event" CCF model, there would be one basic event for each single failure and a basic event for the common-cause failures. This common-cause basic event corresponds to the common-cause failure of all three components as well as combinations of the common-cause failure of two components and the independent failure of the third.

The recommended treatment for a "single event" common-cause model is simply to add the FV of this single event to the independent FV values within the MSPI equation.

Step 4.2: Split Event

Sometimes the common-cause failure of components is addressed at the subcomponent level. For example, common cause for a motor-drive pump can be considered for the motor and the pump. This may be appropriate in that the motor-driven pump could be in a three-pump system with two motor-driven and one turbine-driven pumps. The pumps may be of similar design and may have the same suction source. Therefore, the pumps would be in a three-component common-cause group and the motors in a group of two.

For the MSPIs, a key objective is to capture the change in CCF probability when the associated total failure rate changes. Failures associated with the driver are typically more dynamic than those associated with the pump. Therefore, the recommended treatment for a "split event" is to use the subcomponent FV associated with the highest CCF probability. In the case of the above example, this would be the CCF associated with the motor. This is also consistent with the component grouping used in the MSPIs.

Step 4.3: Multiple Events

Some PRA practitioners and/or PRA software use multiple events to model the impact of CCF. For example, consider a system with three redundant components (and a 1 of 3 success criterion). In a "multiple events" CCF model, there would be a basic event for each single failure, three events, and several events for the common-cause failures. These events would include a basic event for the common-cause failure of all three components and additional basic events for each combination of the common-cause failure of two components and the independent failure of the third. For this example there are a total of four common-cause basic events. However, the number of combinations varies with the success criteria and the degree of redundancy within a given system.

The recommended treatment of "multiple events" is to either use a group FV (if available) to obtain the total FV for all of the common-cause events or, if the group FV cannot be evaluated, to simply add the FVs. The simple addition of the FVs could result in some double counting (overestimation) in the rare case where multiple common-cause basic events for the same common-cause group appear in the same cutset.

Step 4.4: Combined Events

The "combined events" approach addresses the consolidation of failures modes (e.g., fail-to-start and fail-to-run) into a single common-cause basic event. The combined event can either be separated (site-specific PRA model is updated) or estimated.

The following approach can be used to estimate the CCF FV. Cutsets that contain combined events are typically similar to those that contain separate events in that the other failure events in these cutsets are the same. This results in the following relationship:

$$\frac{FV_{FTS}}{UR_{FTS}} \cong \frac{FV_{FTR}}{UR_{FTR}} \cong \frac{FV_{Combined}}{UR_{Combined}}$$

Therefore, the FV for the failure mode of interest can be obtained by determining its contribution to the combined UR CCF value and then multiplying this value by the FV/UR combined value. For example, if failure-to-start is the failure mode of interest, FV_{FTS} would be determined as follows:

$$FV_{FTS} = \frac{FV_{Combined}}{UR_{Combined}} \times UR_{FTS}$$

where UR_{FTS} is the portion of $UR_{Combined}$ associated with the failure to start. Note that UR_{FTS} would be determined based on the examination of the bases for $UR_{Combined}$ in order to determine that portion of the combined event that is associated with the failure to start.

Step 4.5: Conditional Split Fractions

Conditional split fractions are sometimes used in the large event tree methodology to model the CCF impact of redundant trains that are represented by separate top events. The CCF importance may be able to be derived by one of the methods described above or may require other techniques. If other techniques are used, their objective should be to achieve the appropriate CCF contribution similar to the methods above.

Step 4.6: Other

In addition to the above CCF approaches, there may be other unique applications of CCF within PRA models or combinations of the above methods. If other techniques are used, approaches analogous to the above should be applicable. The objective is to reflect the appropriate CCF contribution to each component's FV/UR, as in the methods above.

Step 5: Determine the CCF FV

Based on the above modeling approaches, determine the CCF FV for each identified common-cause group.

Step 6: Add CCF FV to the independent FV values

Add the CCF FV for each common-cause group to the associated independent FV.

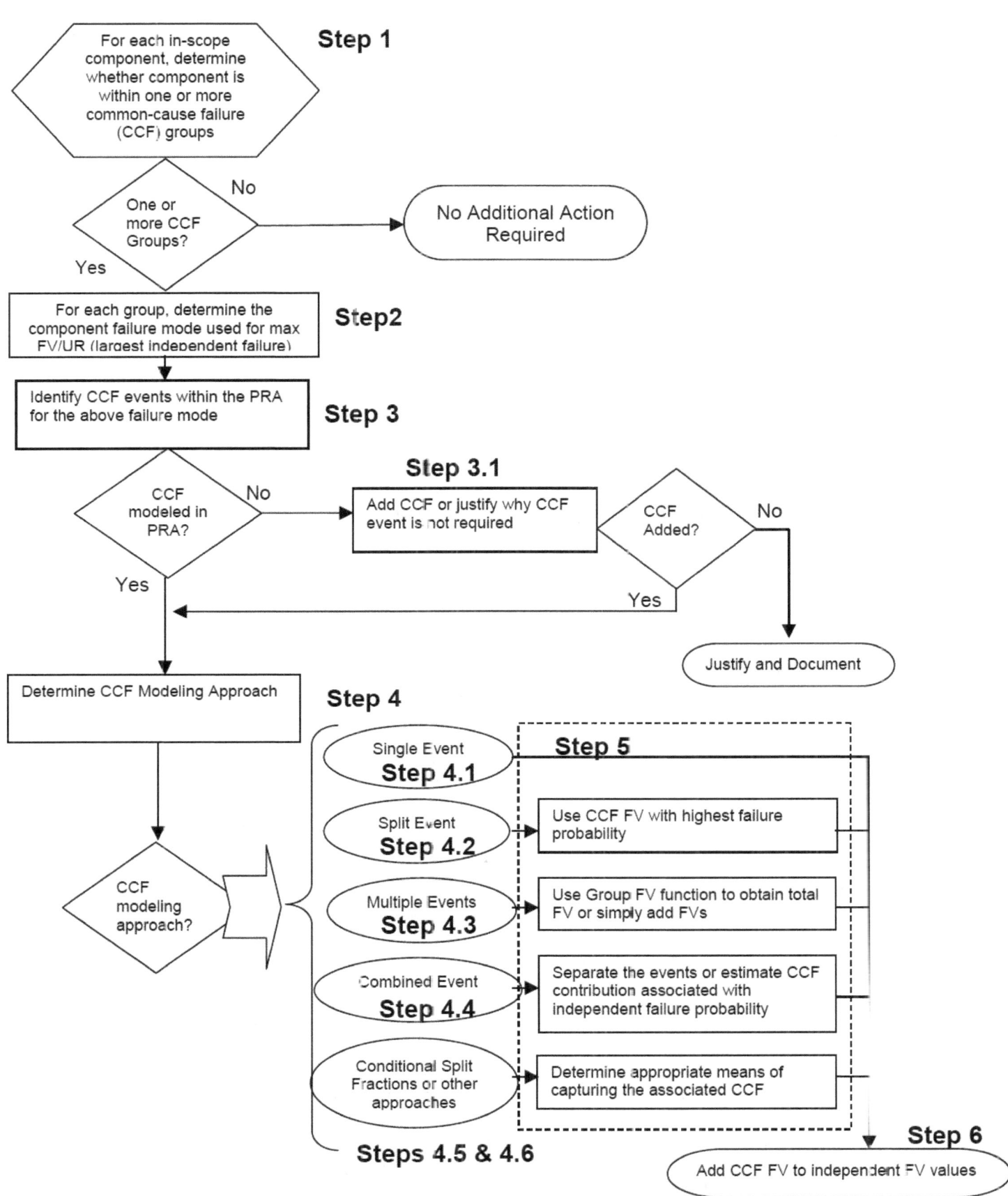

Figure F.1 Evaluation of Common-Cause Failure Contribution to MSPIs

F-11

F.4 Alternate Approach Using Generic Multipliers on FV

Exercises performed by a number of pilot plant participants at the NEI workshop on August 20, 2003, indicated that detailed guidance and training would be required to implement the proposed inclusion of Fussell-Vesely importances for CCF. The exercises also revealed that, in some instances, common-cause modeling includes a complicated coupling of pumps, motors, breakers, and other components. Thus, participants found it difficult to determine the CCF-related FV importances. As a result, the RES staff has provided an alternative approach to address CCF. This alternative approach allows the use of generic multipliers on the FV from independent failures as an appropriate adjustment to account for the effect of CCF.

One simple way to incorporate the impacts of CCF modeling on component Fussell-Vesely importance measures is to apply a CCF multiplier to the importance measure. For a system with two parallel components and system success defined as success of either of the components, the risk model includes three events (independent failure of component 1, independent failure of component 2, and CCF of components 1 and 2). Each of these events has an associated Fussell-Vesely importance factor (FV_1, FV_2, and $FV_{12,CCF}$). To determine the CCF multiplier for a particular component type in a particular system, the Fussell-Vesely importances of all three events are summed and divided by the sum of the Fussell-Vesely importances of the two independent failure events. In equation form, the CCF multiplier is as follows:

$$CCF\ multiplier\ =\ (FV_1 + FV_2 + FV_{12,CCF})/(FV_1 + FV_2)$$

This *CCF multiplier* then can be applied to each of the independent failure event Fussell-Vesely importances. In the example above, FV_1 would be replaced by FV_1*CCF multiplier*, and FV_2 would be replaced by FV_2*CCF multiplier*. This is valid even if the importances of the two components are not equal. For a system with n components, the CCF multiplier would be determined similar to above, but with "$FV_1 + FV_2$" replaced by "$FV_1 + \ldots + FV_n$".

To illustrate how to develop a set of recommended generic CCF multipliers, a two-step process was used. First, the eleven Standardized Plant Analysis Risk (SPAR) resolution models (covering the 20 pilot plants) were used to identify system/component/failure mode CCF multipliers for each model. Results of that effort are presented in Table F.2. Blanks in the table indicate that either the plant has only one such component (and CCF is therefore not applicable) or the SPAR model did not include a CCF event for such components. In a few cases, the data sets were augmented with data from non-pilot plants for better sampling. Then the results in Table F.2 were analyzed to generate a limited set of generic CCF multipliers applicable to the 20 MSPI plants. The generic CCF multipliers are presented in Tables F.3 and F.4. (Table F.4 lists the results by pilot plant rather than by number of components and success criterion.)

The reduced set of generic CCF multipliers in Table F.3 was generated by reviewing the individual plant results in Table F.2. Each table entry was characterized by the number of components modeled, the system success criterion, and other factors such as the availability of backup systems to perform the same function. Then this information was used to group plants with similar CCF multipliers, and a geometric average from those plants was used as the generic CCF multiplier. Also, these multipliers were rounded to 1.25, 1.50, 2.0, 3.0, or 5.0. Finally, for pumps and emergency diesel generators, results for failure-to-start, failure-to-load/run, and failure-to-run were combined to obtain results applicable to all failure modes. If this approach is to be applicable for all 103 plants, a similar process should be used to generate recommended generic CCF multipliers. This expanded effort would include a review of system configurations for all plants and a broader review of SPAR CCF multipliers to ensure applicability to all plants.

Sensitivity studies were performed to assess the effect of generic CCF multipliers on overall MSPI results. The results of these studies were compared to the MSPI values generated for the 100 systems as shown in Tables A.3 and A.4 for 4th quarter 2002.

On a case-by-case basis, the effect of using generic CCF multipliers could be to either *increase* or *decrease* the MSPI results depending on system performance. The CCF multiplier has the effect of increasing the Birnbaum value or coefficient as shown in Equation F.9, for example. If component reliability is worse than baseline, its contribution to URI would be positive, and the larger coefficient resulting from the adjustment for CCF would tend to make these terms more positive. Likewise, terms where performance is better than baseline (negative), would become more negative. In the aggregate, systems with lower MSPI because of the CCF effect would be balanced by systems with higher MSPI owing to CCF.

But in general, the use of generic CCF multipliers is found to increase the number of WHITE MSPI indications, especially where the system MSPI without CCF is a high GREEN and on the margin of the GREEN/WHITE threshold. The results are consistent with numerical simulation that indicates the inclusion of CCF could result in about one-third more WHITE indicators than without accounting for CCF.

F.5   References

F.1   Nuclear Energy Institute (NEI). NEI 99-02 (Draft Report), "Regulatory Assessment Performance Indicator Guideline," Section 2.2 ("Mitigating System Performance Index") and Appendix F ("Methodologies for Computing the Unavailability Index, the Unreliability Index, and Determining Performance Index Validity"). NEI: Washington, DC. 2002.

F.2   A. Mosleh, et al., NUREG/CR-5485, Guidelines on Modeling Common-Cause Failures in Probabilistic Risk Assessment, November 1998.

Table F.2 CCF Multipliers from SPAR Resolution Models

System	Component	Failure Mode	Braidwood	Hope Creek	Limerick	Millstone 2	Millstone 3	Palo Verde	Prairie Island	Salem	San Onofre	South Texas	Surry	Geometric Average
EAC	EDG	FTS	1.41	1.84	6.93	1.11	1.06	1.21	1.11	1.10	1.22	2.31	1.06	1.51
		FTLR												
		FTR	1.51	1.28	1.00	1.11	1.08	1.27	1.14	1.14	1.29	1.97	1.07	1.24
	AOV	FTO/C							1.10					1.10
HPI	MDP Running	FTS	1.12				1.00			1.50			1.25	1.20
		FTR	2.41				1.21			1.23			8.76	2.37
	MDP Standby	FTS	1.18			1.02	1.31	5.88	6.04	3.22	1.27	8.88		2.59
		FTR	1.78			1.00	1.15	3.93	3.38	2.05	1.27	10.72		2.29
	MOV	FTO/C	2.11			1.55	1.04	1.36	1.43	1.26	1.50	1.90	5.50	1.72
	AOV	FTO/C												
HRS	MDP Standby	FTS				1.05	1.74	1.85		2.11	1.00	6.02	1.69	1.84
		FTR				1.00	2.07	1.00		1.15	1.00	2.55	1.09	1.31
	TDP	FTS												
		FTR												
	DDP	FTS												
		FTR												
	MOV	FTO/C				5.74				1.01	1.01	1.81	7.55	2.40
	AOV	FTO/C												2.41
RHR	MDP Standby	FTS	1.61	1.65	1.00	2.40	2.97	1.59	1.13	1.56	1.11	1.56	3.50	1.69
		FTR	1.61	1.29	1.00	2.40	1.90	1.36	1.06	1.27	1.11	1.67	2.17	1.47
	MOV	FTO/C	1.22	2.07	1.01	2.03	1.36	1.31	1.54	1.18	1.50	1.90	14.30	1.81
	AOV	FTO/C												
SWS	MDP Running	FTS	1.33	1.92		1.11	1.24			3.46	2.38	1.82	1.26	1.69
		FTR	7.97	4.39		1.18	51.72			4.81	2.38	20.67	4.21	6.17
	MDP Standby	FTS			1.00			1.14						1.07
		FTR			4.14			1.06						2.10
	DDP	FTS							1.25				1.00	1.12
		FTR							1.80				1.00	1.34
	MOV	FTO/C				1.09	1.24					6.31	1.13	2.07
	AOV	FTO/C									1.07			1.08
CCW	MDP Running	FTS	1.30			1.24			1.07	1.49	1.39	1.93	1.10	1.34
		FTR	1.76			1.15			1.69	1.94	1.39	6.67	1.43	1.90
	MDP Standby	FTS						2.59						2.59
		FTR						1.98						1.98
	MOV	FTO/C												
	AOV	FTO/C				1.54						3.28		2.24

1.00 = Truncated (or calculated) CCF

Blank = single component, components don't exist, or components and/or CCF not modeled

All MOVs	1.86
All AOVs	1.68
All MDPs Running	2.17
All MDPs Standby	1.79
All EDGs	1.37

Acronyms: AOV (air-operated valve), CCF (common-cause failure), CCW (component cooling water), DDP (diesel-driven pump), EAC (emergency ac power), EDG (emergency diesel generator), FTLR (fail to load and run for 1 hour), FTO/C (fail to open or close), FTR (fail to run), FTS (fail to start), HPI (high pressure injection system), HRS (heat removal system), MDP (motor-driven pump), MOV (motor-operated valve), RHR (residual heat removal system), SWS (service water system), TDP (turbine-driven pump)

Table F.3 Sample Generic CCF Multipliers

System	Component	Generic CCF Multiplier					Comments
		1.25	1.50	2.00	3.00	5.00	
EAC	EDG	2 EDGs (1/2) or 3 EDGs (2/3)	4 EDGs (1/4) with other diverse sources of power	3 EDGs (1/3)		4 EDGs (1/4) and no diverse sources of power	4 EDG case (with no diverse sources of power) includes information from SPAR Rev. 3 models for Browns Ferry 3 and Fitzpatrick.
HPI	MDP Running		With SI and CVC		With only CVC		
	MDP Standby		With SI and CVC		With only SI		
HRS	MDP Standby	2 MDPs (1/2)			3 MDPs (1/3)		Information from SPAR Rev. 3 models for Calvert Cliffs, Davis Besse and Turkey Point used.
	TDP	2 TDPs and 1 MDP			3 TDPs and no MDPs		
RHR	MDP Standby		All				
SWS	MDP Running				All		
	MDP Standby		All				
	DDP	All					
CCW	MDP Running		All				
	MDP Standby			All			
All	MOV			All			
All	AOV		All				

Note - Success criterion indicated in parentheses.

Note - Generic CCF multipliers obtained from SPAR resolution model results for 11 pilot plants, unless otherwise indicated.

Acronyms: AOV (air-operated valve), CCF (common-cause failure), CCW (component cooling water), CVC (chemical and volume control system), DDP (diesel-driven pump), EAC (emergency ac power), EDG (emergency diesel generator), HPI (high pressure injection system), HRS (heat removal system), MDP (motor-driven pump), MOV (motor-operated valve), RHR (residual heat removal system), SI (safety injection system), SWS (service water system), TDP (turbine-driven pump)

Table F.4 Sample Generic CCF Multipliers by Pilot Plant

System	Component	Generic CCF Multiplier					Comments
		1.25	1.50	2.00	3.00	5.00	
EAC	EDG	Braidwood, Millstone 2, Millstone 3, Palo Verde, Prairie Island, Salem, San Onofre, Surry	Hope Creek	South Texas		Limerick	4 EDG case (with no diverse sources of power) includes information from SPAR Rev. 3 models for Browns Ferry 3 and Fitzpatrick.
HPI	MDP Running		Braidwood, Millstone 3, Salem		Surry		
	MDP Standby		Braidwood, Millstone 3, Salem		Millstone 2, Palo Verde, Prairie Island, San Onofre, South Texas		
HRS	MDP Standby	Millstone 2, Millstone 3, Palo Verde, Salem, San Onofre, Surry			South Texas		
	TDP	No MSPI pilot plants			No MSPI pilot plants		Information from SPAR Rev. 3 models for Calvert Cliffs, Davis Besse and Turkey Point used.
RHR	MDP Standby		All				
SWS	MDP Running				All		
	MDP Standby		All				
	DDP	All					
CCW	MDP Running		All				
	MDP Standby			All			
All	MOV			All			
All	AOV		All				

Note - Success criterion indicated in parentheses.

Note - Generic CCF multipliers obtained from SPAR resolution model results for 11 pilot plants, unless otherwise indicated.

Acronyms: AOV (air-operated valve), CCF (common-cause failure), CCW (component cooling water), DDP (diesel-driven pump), EAC (emergency ac power), EDG (emergency diesel generator), HPI (high pressure injection system), HRS (heat removal system), MDP (motor-driven pump), MOV (motor-operated valve), MSPI (mitigating systems performance index), RHR (residual heat removal system), SWS (service water system), TDP (turbine-driven pump)

APPENDIX G. TECHNICAL BASIS FOR EXCLUDING ACTIVE VALVES BASED ON BIRNBAUM IMPORTANCE

Appendix G
Technical Basis for Excluding Active
Valves Based on Birnbaum Importance

G.1   Background

Appendix F of Draft NEI 99-02 MSPI Rev 0 provides clarifying notes as to the criteria for determining those components that should be monitored. For example, all pumps and diesel-generators are included in the performance index. Specific guidance is provided on page F-9 for valves, whether in series or parallel for multi-train systems. The guidance is prescriptive in nature and is intended to ensure to a first order of approximation that important valves within a system are included.

The expectation is that the number of valves to be monitored should not be too different from the number of pumps in the system. Thus, in a three-train system consisting of three pumps, one should expect the number of valves to be monitored to be on the order of two to six. Certainly 10 or more valves to be monitored within a system should be the rare exception.

For the 20 pilot plants in the program, the average number of components for all 6 systems combined has been found to be fewer than 50, as follows:

- about 16 pumps
- about 24 valves
- 2 to 4 emergency diesel-generators
- the occasional circuit breaker for electrical cross-tie

The above counts meet general expectations. However, there are instances where, for several reasons, the number of valves to be monitored in total has been determined to be as high as 46. This far exceeds expectations and can pose a large data collection burden, with no clear benefit in return.

G.2   Birnbaum Cutoff for Excluding Valves

Based on an analysis of all of the valves monitored by the 20 pilot plants, it is possible to exclude low importance valves without affecting the overall results of the MSPI. The analysis considered both FV/UR and Birnbaum (CDF * FV/UR) as possible criterion for excluding active valves from the MSPI. Birnbaum has been deemed to be more appropriate since it is the measure directly used in the calculation of URI, and URI is the figure-of-merit of interest here.

Figure G.1 shows the average number of active valves (mainly air-operated and motor-operated) per nuclear unit that would be monitored as a function of possible cutoff in Birnbaum, based on pilot plant results. Lowest and highest valve counts are also shown for comparison. As the Birnbaum increases, there is a large initial drop in the average valve count, owing to a clustering of low importance valves. The plot flattens out considerably after 1×10^{-6}/yr or so. Clearly, there is diminishing return after about 1×10^{-6}/yr.

Figure G.2 shows the potential unaccounted for delta URI that could arise from the exclusion of low importance valves from the MSPI. The analysis is conservative because it assumes that the excluded valves in question each could have had three failures over 3 years. The potentially unaccounted for delta URI plot remains flat for the "average" case through 1×10^{-6}/yr, before increasing slowly thereafter. The unaccounted for delta URI for the highest valve count plant is conservatively calculated as being about 1×10^{-7}/yr at a Birnbaum of 1×10^{-6}/yr. This unaccounted for delta URI is only 10% of the value necessary to turn the indicator for the system WHITE. And this assessment is additionally conservative because not all valves that would be excluded would necessarily be in the same system.

In consideration of the benefits to be gained by excluding low risk valves, and the insignificant impact on MSPI results, the exclusion of active valves with Birnbaum of less than 1×10^{-6}/yr is appropriate. ***Based on the discussion below, the common-cause contribution to FV (and Birnbaum) must be added to the valve Birnbaums before the cutoff is applied.***

G.3   Other Considerations

Appendix F discusses the need to include the common-cause contribution to FV in the overall approach to the MSPI. Since Figure G.1 does *not* include the adjustment to valve Birnbaums owing to common cause, the potential benefit in terms of the number of valves excluded from scope could be somewhat less than shown. The effect of including the adjustment to FV for common cause would be to shift the three plots in Figure G.1 to the right. Without having available the FV due to common cause from the plant PRA for all the pilot plant valves, the exact effect can not be ascertained. If the option of using *generic multipliers* as discussed in Appendix F were used, then the impact could be estimated. For motor-operated valves, a generic multiplier of 2.0 has been recommended. This would effectively reduce the unadjusted Birnbaum cutoff (i.e., without common cause) from 1×10^{-6}/yr to 5×10^{-7}/yr. Figure G.1 shows that using a Birnbaum cutoff of 1×10^{-6}/yr reduces the average number of valves per plant from 24 to about 17. If common cause using the generic multiplier is included, this average is estimated to be 18 instead.

Another important consideration is whether or not some minimum number of valves should remain in-scope regardless of their risk importance. *Any valves that meet the cutoff criterion of 1×10^{-6}/yr on Birnbaum (including common cause) do not impact URI in any way.* However, there could be undesirable consequences of monitoring too few valves in MSPI. For one, the more valves that are monitored, the larger the pool of similar valves and the higher the number of demands. If a larger population is considered, the URI is less sensitive to small numbers of failures of valves, and less likely to result in a false WHITE for a small (statistically not unlikely) number of failures. Secondly, valves not monitored in the MSPI could be subject to the inspection process. Thirdly, as the plant PRA model changes owing to changes in plant design or equipment performance, it is likely that importance measures also change. A valve with a Birnbaum just under the 1×10^{-6}/yr cutoff probably should be included because of its potential to meet the criterion at some future point. It therefore seems reasonable to ensure a minimum number of valves within fluid systems are monitored by the MSPI, regardless of their risk significance.

The next logical question would be to ask whether such a cutoff could be applied to components other than valves. Analysis was performed for pumps in a way similar to valves. A case could have been made to exclude some pumps based strictly on risk. However, since the pumps are at the core of the system reliability, it would be inconsistent with the intent of the MSPI to exclude pumps from monitoring.

G.4 Process

The approach to address the option regarding which valves to monitor in the MSPI should proceed as follows:

(1) Identify all active valves that meet the prescriptive criteria per NEI 99-02.

(2) Calculate the independent Fussell-Vesely (FV) importance for all valves in (1).

(3) Calculate the common-cause contribution to FV for all valves in (1) per Appendix F in this report. Apportion the FV due to common cause and add them to FV for independent failures. (For example, if FV_{ccf} were 0.02 for a two-valve configuration, 0.01 would be added to the FV for each independent failure.)

(4) Calculate the Birnbaum (= CDF * FV/UR) for all valves. (If the option to use generic multipliers for common cause is invoked, the effective Birnbaum cutoff would be the unadjusted B divided by the generic multiplier, rather than 1×10^{-6}/yr.)

(5) Identify which valves are required to be monitored (B > 1×10^{-6}/yr), and those that are optional (B < 1×10^{-6}/yr.)

(6) Based on a consideration of the (a) potential data collection burden if the list of valves is large, (b) the desirability of having a large enough pool of valves, and (c) the margin of valves from the 1×10^{-6}/yr cutoff, clearly identify the list of valves that are to be monitored for the duration of the indicator.

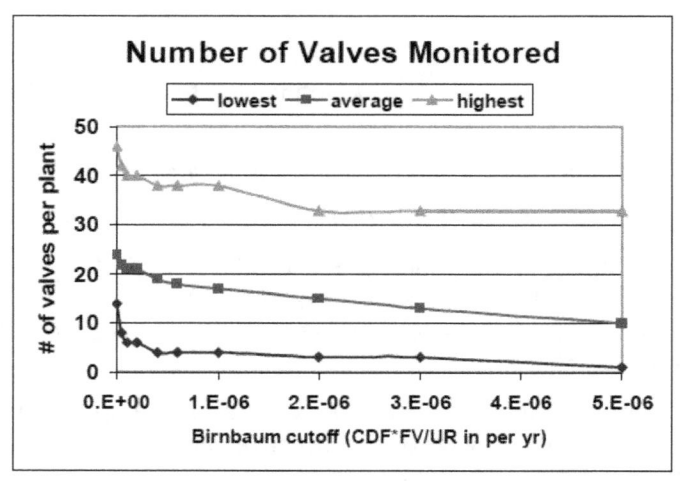

Figure G.1 Average Number of Values Monitored

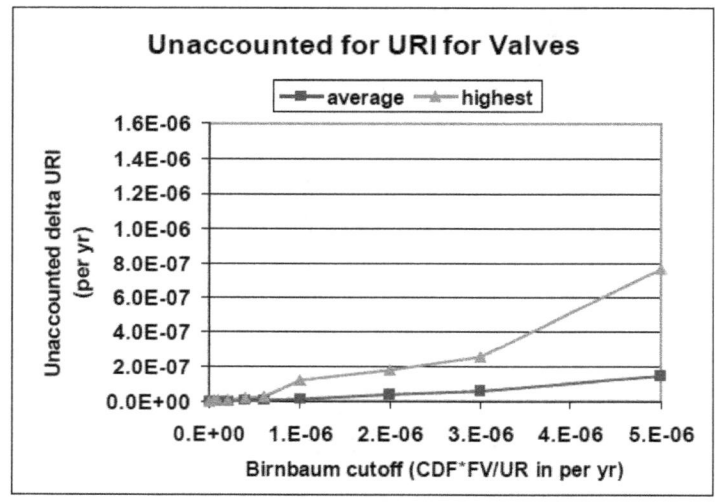

Figure G.2 Estimate of Unaccounted for URI for Valves

APPENDIX H. TECHNICAL BASIS FOR INCLUDING THE CONTRIBUTION OF SUPPORT SYSTEM INITIATORS TO FUSSELL-VESELY IMPORTANCE

Appendix H
Technical Basis for Including the
Contribution of Support System Initiators to Fussell-Vesely Importance

H.1 Background

The MSPI is calculated for five indicators consisting of six systems. Two of the six systems, component cooling water and service water, are to be combined into one indicator called the "cooling water support system." A primary reason for combining these systems into one indicator is owing to the large variability in design from plant-to-plant. For the majority of nuclear plants, the cooling water systems provide cooling to secondary and auxiliary systems such as the turbine-generator, as well as to safety systems such as the emergency diesel-generators and residual heat removal heat exchangers. However, in a number of plants, cooling water systems have been separated into those that provide cooling strictly to non-safety components and those that cool only safety systems. Still other plants utilize safety-related service water to directly cool safety-related systems, and do not have the intermediate safety-related component cooling water system. In the MSPI, only those cooling water support systems that have some safety-related function (and are not dedicated to cooling just one component) are to be included.

Support systems such as service water contribute to a plant PRA model in two ways. First, the service water system provides a "support" role whereby it cools other support systems such as emergency diesel generators or even "frontline" systems, depending on the design. These are modeled appropriately in the PRA through the use of linked fault trees or large event trees. Second, if the loss of the cooling water system such as service water could also result in a plant transient, automatic scram, or is likely to lead to a manual scram, then that system is also modeled as a potential initiating event in the PRA. Thus, a component such as a service water pump could impact the overall plant PRA results (a) because of its function in cooling needed equipment *following* a transient, and (b) through the potential to *initiate* a plant transient. Of all the systems within scope of the MSPI, the service water system (SWS) and component cooling water (CCW) system are the two that could serve in the dual roles of both supporting other systems when called upon, and initiating a transient if the SWS or CCW is lost entirely or substantially degraded.

The plant PRA models calculate various risk measures such as Fussell-Vesely importance, risk achievement worth (RAW) and Birnbaum importance from basic event probabilities. All PRA models can provide such risk measures for SWS and CCW components of interest in the MSPI. However, while all the models include the component's contribution from the "support system" role of the SWS and CCW, not all models include the contribution to the importance measures from the loss of SWS or CCW as an initiating event. This is because the initiating event frequencies used in some plant PRA have been based on plant and/or industry experience, and use an explicit value for the frequency. The frequency may use a distribution with mean and variance, but the value has been calculated is some way separate from the linked PRA model. In other models, the PRA analyst may have chosen to link a loss of SWS initiator fault tree directly into the computer model of the PRA. This is a matter of practice and convenience that is left to the discretion of the analyst. The Standardized Plant Analysis Risk (SPAR) models, for example, use initiating event frequencies for loss of SWS and loss of CCW that are based largely on industry experience. SPAR does not use fault trees for these initiators, but could be changed to do so. Either approach is acceptable so long as it is based on valid equipment performance data, takes into account the potential for common mode failure based on plant-specific characteristics and design, properly conditions mitigating system failure on initiating event characteristics, and is generally consistent with industry operating experience.

H.2   Contribution to Fussell-Vesely Importance Measure

All other things being equal, a plant PRA model that uses initiator fault trees explicitly for loss of SWS and/or CCW (where Importance of the initiating event components is accounted for) will result in higher Fussell-Vesely (FV) and Birnbaum risk measures for an associated basic event than a model that uses a point-estimate frequency. The difference between the two approaches would be a function of the importance of that initiator to the overall calculated core damage frequency (CDF), as well as the importance of the particular component (and basic event) within the SWS or CCW of interest. During the MSPI workshop on January 21, 2003, a survey was taken of the pilot plant participants. Plant PRA models fell into three categories, including (a) those that used fault trees for loss of SWS and loss of CCW initiators that were directly linked in the PRA model, (b) those that used fault trees and/or event trees outside of the linked PRA model to quantify the frequencies, which were manually entered into the PRA model no differently than a medium loss-of-coolant accident (LOCA) frequency, and (c) those that used frequencies based on industry experience, updated with plant-specific data. Category "a" is the most prevalent, with about two-thirds of the pilot plants using this approach. These differences in approach clearly result in an inconsistency for the purpose of the MSPI; the MSPI methodology relies heavily on using calculated risk measures (FV divided by basic event probability) rather than (say) a re-quantification of the entire PRA model.

Given this inconsistency, there are three options for consideration:

(1) For those plant PRAs that have used linked SWS and CCW initiator fault trees, require that they substitute point-estimate frequencies in lieu of using the linked trees.

(2) For those plant PRAs that have used point-estimate frequencies for loss of SWS and CCW, ensure that they account for the contribution of the SWS and CCW initiators in the FV computation for the components within scope of the MSPI.

(3) Ignore the inconsistent approaches.

Sensitivity studies have been performed by some pilot plant analysts to identify the importance of including the contribution of support system initiators to the FV risk measure. Calculations were performed first by using the existing linked fault tree initiator models, and next with the fault tree initiator essentially turned off. Differences in FV using the two approaches can be expected to be strong functions of the following factors:

• the importance of the initiator to overall CDF
• importance of the component within the system
• system configuration and design
• importance of recovery actions and success criteria

At the lower end, the differences in calculated FV with and without initiator fault trees were shown to be less than one percent. At the upper end, differences as high as an order of magnitude in FV were seen for some components. Clearly, the potentially significant contribution of support system initiators to FV rules out options "1" and "3". The only viable option is "2", that is, to account for the contribution of the support system initiator to FV. Some have argued that in a mitigating system performance index, these contributions of _initiators_ should not be included at all. But the loss SWS or CCW initiators cascading to core damage also implies that these components would not have been available to support their mitigation function as well. The contribution of SWS and CCW components to FV, both as initiators and mitigators, need to be included if the full risk importance is to be properly accounted for.

H.3   Process to Account for Support System Initiators

Figure H.1 shows the process to account for the contribution of support system initiators to FV. Clearly, if the safety-related CCW and/or SWS to be monitored in the MSPI are strictly standby systems, then their loss can not initiate a plant transient. The calculated FV values for the CCW/SWS components are proper and no further action is necessary.

In the second diamond, if initiator fault trees are being used, then the contribution of initiators to FV is accounted for. *However, it is critical that the same basic event ID is used both in the support system modeling and in the initiator fault tree.* FV importance is calculated on a *basic event* level, and the use of different IDs would result in the full contribution of a failure mode to FV not being captured. This would necessitate adding the contributions manually.

If different basic event probabilities UR_c and UR_{ie} are used because of different mission times for the same component failure mode, addition of the FV for the support system aspect to the FV contribution from the initiator fault tree would give consistent and correct results. In theory, it is the Birnbaums (= CDF * FV/UR) that are directly additive. But in the fundamental expression for URI shown below, if UR is proportional to the mission time via a fail-to-run expression λT, then the increase in the denominator is cancelled by the increase in the term in the parenthesis. Birnbaum is preserved by adding the FV values in this situation.

$$URI = CDF_P \sum_{j=1}^{n_c} \left[\frac{FV_{UR_{cj}}}{UR_{pcj}} \right]_{max} (UR_{Bcj} - UR_{BLcj})$$

Assuming that no initiator fault trees exist, it is possible to avoid the need to include the contribution of initiators to FV, as shown in the third diamond of Figure H.1. Analysis of pumps and valves indicate that a component with a Birnbaum of 1×10^{-6}/yr typically contributes of the order of 1×10^{-9} to 1×10^{-8}/yr to delta URI. Even if the inclusion of the contribution of the initiator to FV could increase the Birnbaum and hence delta URI by an order of magnitude, it still would make the component a relatively insignificant contributor to the overall system MSPI. Hence, if all CCW/SWS components to be monitored in the MSPI have their Birnbaum (maximum for all failure modes) less than 1×10^{-6}/yr, then it is not necessary to take further action. Only if none of the above conditions are met is it necessary to account for the contribution of initiators to FV.

In the proposed resolution (the rectangle of Figure H.1), licensees would be given two options. Those plant PRA models that do not use fault trees for loss of service water and/or loss of component cooling water could either a) add such fault trees and recalculate the FV importance measures, or b) use an approximation that adjusts the FV to account for the contribution in a way proportional to the importance of the system initiator to core damage frequency, and proportional to the importance of the component within the system. Presumably, if numerous components within CCW and/or SWS are impacted, creating new initiator fault trees may well be the preferred way to proceed. In this process, care must be taken to account for all basic events associated with a component since the identifiers for these events could be different between the initiating event fault tree and the mitigating system fault tree. The fault trees would have to adequately include the potential contribution from common-cause events, as seen through industry operating experience.

H.4   Alternate Approach to Calculate FV for Support System Initiators

Now presume that only two components in the CCW or SWS are shown to have Birnbaums greater than 1×10^{-6}/yr. Why should it be necessary to create entirely new initiator fault trees when most of the components would have no impact on the calculated system URI (and MSPI)? Since the MSPI algorithm relies only on inputted FV/UR, an adjustment to two FVs is all that is called for. As discussed above, the adjustment is based on the following factors:

- the proportionality of the importance of the system initiator to CDF
- the proportionality of the importance of the component to the system

Mathematically,

- Let FV_{ie} be the Fussell-Vesely contribution for the initiating event in question (e.g., loss of service water).

- Let FV_{sc} be the Fussell-Vesely *within the system fault tree only* for component c (i.e., the ratio of the sum of the cut sets contribution in which that component appears to the overall system failure probability).

- Let FV_c be the Fussell-Vesely for CDF for component c as calculated from the PRA model. This does not include any contribution from initiating events.

The adjusted FV to include in the MSPI is then

$$FV_c + [\, FV_{ie} * FV_{sc} \,] \qquad\qquad\qquad \text{(Eq. H.1)}$$

To assess the accuracy of this approximation, several licensees compared the adjusted FV for a dozen or so SWS and CCW components to the correct FV as computed within the PRA model. The results are provided Figure H.2. This adjustment is shown to be conservative, yielding from zero to approximately 25% higher FV (based on regression analysis) than would be expected using an initiator fault tree. These differences in results arise because of differences in success criteria and recovery actions in the initiator tree, whereas less credit is often given in the support system fault tree model. Hence, the approach is conservative. Given this potential conservatism in the approximation to adjust the FV, licensees may well choose to develop initiator fault trees for loss of service water and loss of component cooling water for the purpose of the MSPI.

Note that the above discussion focused on the need to account for the effect of support system initiators on FV for basic event probabilities related to component unreliabilities. Since train unavailabilities can also contribute to initiator frequency, a similar adjustment would be necessary for the FV for train unavailability if initiator fault trees are not used.

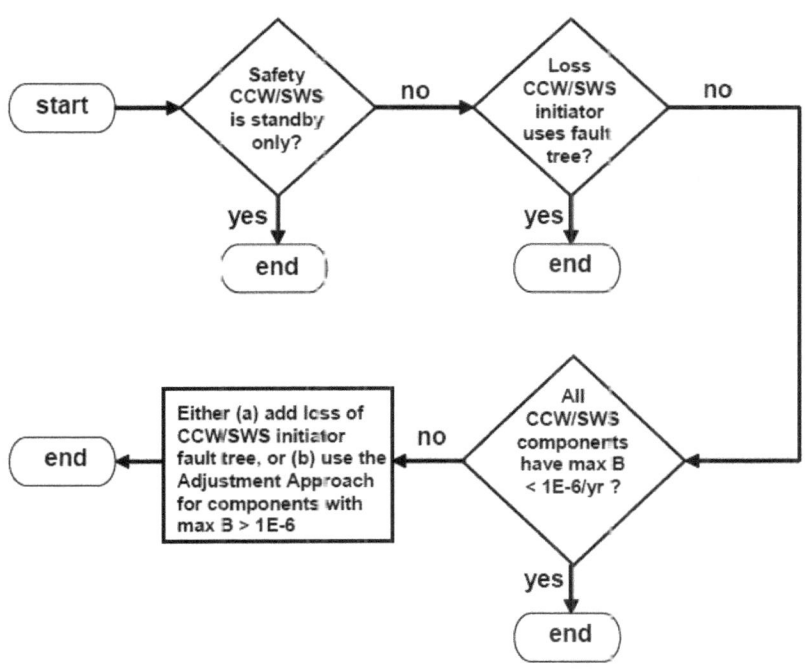

Figure H.1 Flow Chart for Support System Initiators

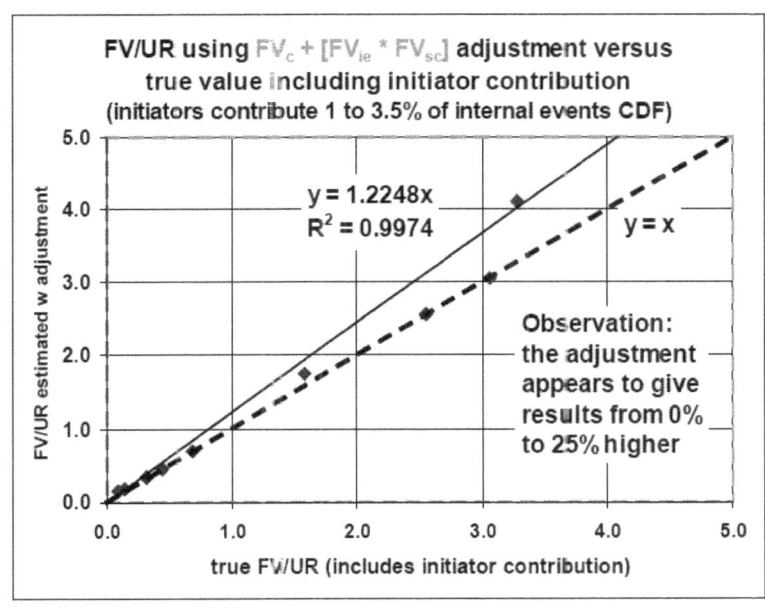

Figure H.2 Comparison of Approximation to Exact Solution

APPENDIX I. MSPI/SSU/SDP BENCHMARK

Appendix I
MSPI/SSU/SDP Benchmark

I.1 Introduction

To assess the characteristics of the Mitigating Systems Performance Index (MSPI), comparisons were made with corresponding Reactor Oversight Process (ROP) safety system unavailability (SSU) indicators and Significance Determination Process (SDP) evaluations to the extent possible. The limitations of this comparison are recognized in that the MSPI and SSU indicators are based on aggregate quarterly measures using rolling 3-year base periods, whereas the SDP evaluations are for single events. The comparisons focus on the performance color predictions (GREEN, WHITE, YELLOW, or RED) from each of these different measures. Two different comparisons were made:

- MSPI indicators for each of the component failures identified in the MSPI pilot program over the period 2000 – 2002 (and corresponding SSU and SDP results)

- SSU and SDP mitigating system non-GREEN evaluations over the period 2000 – 2002 (and corresponding MSPI results)

I.2 Sources of Information

For the MSPI data and results, spreadsheets based on an NEI template were submitted by each of the MSPI pilot plants. There were a total of 77 component failures over the 3-year period from 2000 through 2002 for monitored components at the 20 pilot plants. The spreadsheets automatically calculate the MSPI delta core damage frequency (ΔCDF) results for each system, given the component performance data and train unavailability data over a 3-year period. These calculations use the plant CDF and component and train Fussell-Vesely importance measures obtained from the plant probabilistic risk assessment (PRA) model. However, the component unreliability baselines built into the spreadsheets were replaced by the "Year 2000" baselines recommended in Appendix C of this report.

SSU performance indicator results were obtained from the Nuclear Regulatory Commission (NRC) Web site (Ref. I.1) under the "Historical Performance for Previous Quarters" page. This included both train unavailability data and the resulting color.

Finally, SDP evaluation information was obtained from similar sources as the SSU. To identify SDP evaluations related to the MSPI component failures, the same NRC website was used. Under the "Historical Performance for Previous Quarters" page, individual plant inspection findings by quarter were reviewed. These findings listed corresponding inspection report numbers. Then the "List of Inspection Reports" page was used to obtain actual inspection reports. These reports were reviewed to see if SDP evaluation results were referenced. The SDP non-GREEN findings over the period 2000 – 2002 were compiled primarily from this information.

I.3   MSPI Component Failure Comparison

The MSPI covers six mitigating systems and five indicators: emergency ac power (EAC), high-pressure injection (HPI), heat removal system (HRS), residual heat removal (RHR), and service water system/component cooling water (SWS/CCW). Within each of these systems, a subset of components is included within the scope of the MSPI, and performance of these components is tracked and reported quarterly. For the period 2000 – 2002 (termed the 4Q2002 data period), the 20 MSPI pilot plants identified 77 failures of monitored components. These failures are listed in Table I.1, along with the quarter in which the failure occurred.

For each MSPI component failure in Table I.1, a corresponding MSPI ΔCDF was determined using the NEI spreadsheet (with the "Year 2000" baselines). The MSPI methodology uses a rolling 3-year period of data in its calculation routine. This implies that if a failure occurred in 4Q2000, then data over the period 1998 – 2000 would normally be used. However, data before 3Q1999 are not available within the spreadsheets. Therefore, for consistency, all of the MSPI calculations presented in Table I.1 were performed using data over the period 2000 – 2002 (4Q2002 results). These data include monitored component performance (failures and demands or hours) and train unavailability hours and required (i.e., reactor critical) hours.

If a system includes more than one failure, then the failures are listed chronologically by quarter in Table I.1. The MSPI calculation for a given system includes all of the component failures down to the one in question. As an example, consider Braidwood 1 HRS in Table I.1. The MSPI calculation for the first failure (DDP FTR, 2Q2001) includes only that failure. However, the MSPI calculation for the second failure (DDP FTS, 4Q2001) includes both failures. Finally, the MSPI calculation for the third failure (DDP FTS, 1Q2002) includes all three failures. This calculation approach mimics the MSPI calculations performed on a quarterly basis, except that for the other components (all with no failures) and trains within the system, the performance data are always based on 2000 – 2002, rather than on a rolling 3-year period.

A special case for multiple component failures in the same system involves several failures occurring within a single quarter. In that case, the MSPI calculation for each of those failures includes all of the failures occurring within that quarter (and also system failures occurring before that quarter). That situation occurs for Hope Creek EAC for 4Q2002, where two emergency diesel generator (EDG) FTSs occurred. The MSPI calculation for each of those two failures includes both EDG FTSs within that quarter, plus the four EDG failures occurring before 4Q2002. Because of several cases with multiple failures within one quarter, the 77 MSPI failures actually correspond to 64 MSPI quarterly indicators.

Finally, there is the potential for a component failure to result in a GREEN MSPI for the quarter in which the failure occurred, and yet result in a WHITE in a succeeding quarter (with no additional component failures) because of larger than expected train unavailability. This is observed with the Hope Creek HPI motor-operated valve (MOV) FTO/C event in 2Q2001. The MSPI for that quarter and successive quarters up through 2Q2002 is GREEN. However, the indication for the next quarter, 3Q2002, results in a WHITE because of a relatively large train unavailability outage during 3Q2002. Although this type of multiple quarter MSPI calculation was not performed formally for all of the 77 MSPI failures, the failures with large unreliability contributions to the MSPI were reviewed to identify quarters with large unavailability contributions. No cases other than the Hope Creek HPI indicator were identified.

MSPI results (ΔCDF and color) for the 77 component failures are presented in Table I.1. Of the 77 cases, 8 result in WHITE indications (including the Hope Creek HPI failure discussed previously), while the remaining 69 are GREEN. In terms of the more meaningful MSPI quarterly indicators, 5 of the 64 quarterly evaluations are WHITE, while 59 are GREEN.

If the proposed *frontstop* outlined in Appendix D of this report were applied, then only 2 failures (out of 77) result in a WHITE color. Specifically, those are Braidwood 1 HRS (DDP FTS, 1Q2002) and Hope Creek HPI (MOV FTC/C 2Q2002 but evaluated through 3Q2002). The other 6 failures with WHITE indications revert to GREEN. In terms of MSPI quarterly calculations, only 2 of 64 are WHITE.

Table I.1 also shows the corresponding SSU results, in terms of the unplanned outage and fault exposure times, the unavailability value (expressed as a percent), and the color. Because the SSU does not include the SWS/CCW, the SSU results are listed as "N/A" for MSPI failures occurring within these systems. There are a total of 55 SSU entries in Table I.1 not labeled as "N/A" (counting only 1 entry for the Hope Creek HPI event). Of these 55 entries, 9 are WHITE and 46 are GREEN. These 55 entries correspond with 47 quarterly indicators, again because of multiple failures occurring within a single quarter. Of the 47 quarterly indicators, 8 are WHITE and 39 are GREEN.

Of the 55 MSPI component failures not occurring in the SWS/CCW, there are 7 cases where the MSPI calculation is GREEN while the SSU is WHITE. Also, there are 6 cases where the MSPI calculation is WHITE while the SSU is GREEN. In terms of the 64 MSPI quarterly indicators, there are 6 cases where the MSPI value is GREEN while the SSU is WHITE, and 3 cases where the MSPI is WHITE while the SSU is GREEN.

Note that if the proposed frontstop were applied, then there are 8 cases where the MSPI calculation is GREEN while the SSU is WHITE. Also, there is 1 case where the MSPI is WHITE and the SSU is GREEN. In terms of the 64 MSPI quarterly indicators, there are 7 cases where the MSPI evaluation is GREEN while the SSU is WHITE, and 1 case where the MSPI is WHITE and the SSU is GREEN.

Finally, Table I.1 shows the corresponding SDP evaluation results that were reported in inspection reports. Of the 77 MSPI failures, SDP evaluations mentioned in the inspection reports covered 16 of the failures. Of these 16, 2 are WHITE. This indicates that overall, the SDP methodology resulted in 2 WHITES and 75 GREENS or no SDP findings for the 77 MSPI failures. This is in comparison to the MSPI calculations, in which 8 of 77 are WHITE. If the front stop is applied, then the MSPI calculations result in 2 WHITES out of 77 MSPI failures.

Comparing individual component failure results, there is one failure where the MSPI is GREEN while the SDP is WHITE (Millstone 2 HRS). Also, there are seven failures where the MSPI is WHITE and the SDP is GREEN (or no SDP finding). If the frontstop is applied, then there are two failures where the MSPI is GREEN and the SDP is WHITE (Millstone 2 HRS and Salem 1 EAC, EDG FTR, 3Q2002), and there are two failures where the MSPI is WHITE and the SDP is GREEN or there is no SDP finding (Braidwood 1 HRS, DDP FTS, 1Q2002 and Hope Creek HPI, MOV failure to open or close (FTO/C), 2Q2002 but also evaluated for 3Q2002).

Finally, the MSPI results in Table I.1 were reviewed to determine whether any color changes might occur if the proposed common-cause failure (CCF) adjustments to component Fussel-Vesely importances were used. These adjustments are discussed in Appendix F of this report.

Including CCF adjustments could change the numerical results in Table I.1, and the quarter in which some indicators become WHITE. But only in one case might the inclusion of CCF affect the overall color outcome (Surry-1 SWS/CCW may become WHITE), and here the case is borderline and dependent on the PRA model used and the CCF method applied.

I.4   SSU and SDP Whites Comparison

The other comparison covers SSU and SDP WHITES identified over the period 2000 – 2002 for the six MSPI systems within the 20 MSPI pilot plants. Only one SSU WHITE during 2000 – 2002 is not listed in Table I.1. That SSU is listed in Table I.2. The SSU WHITE at Millstone-2 for HPI (3Q2000) was the result of a component condition identified during periodic testing. However, an actual failure did not occur during testing. Therefore, the MSPI is not applicable for this event. The SDP evaluation for this event resulted in GREEN, as noted in Table I.2.

As indicated in Section I.2, the SDP WHITES were identified by a review of SDP findings for 2000 – 2002. Overall, there were six SDP WHITES identified. Two are listed in Table I.1. However, four of these SDPs cover component failures or discovered conditions that are outside the scope of the MSPI. These four events are listed in Table I.2. For these cases, only the SDP would be used for assessing their safety significance per the guidelines. Of the remaining two events (listed in Table I.1), the MSPI results without the frontstop are WHITE (agreeing with the SDP) for the Salem 1 EAC EDG FTR event and GREEN for the other (Millstone 2 HRS). With the frontstop applied, both of the MSPI results are GREEN. For the SSU, two of the six events are not applicable because they cover the SWS/CCW (not explicitly within the scope of the SSU). Of the remaining four events, the SSU results include two WHITES (agreeing with the SDP) and two GREENS.

I.5   Summary of Comparisons

During the 3-year period from 2000 through 2002, 77 MSPI component failures occurred, corresponding to 64 quarterly MSPI indicators (because of multiple failures occurring within a single quarter). For these 77 failures, the MSPI calculations result in 8 WHITES and 69 GREENS. If the proposed frontstop were used, there would be 2 WHITES and 75 GREENS. In terms of the 64 quarterly MSPI indicators, 5 result in WHITE, while 59 are GREEN. With the proposed frontstop, 2 are WHITE and 62 are GREEN. *However, because some WHITE MSPI indicators remain so for more than one quarter, the number of unique WHITE MSPI indicators for the 20 pilot plants over the 3 years is 4 without the frontstop and 2 with the frontstop.* The unique number of WHITES is important to note because increased regulatory attention would probably occur only once for consecutive quarterly WHITE indicators on the same system.

For the SSU, 55 of the MSPI component failures are applicable (excluding SWS/CCW failures). The SSU results for these 55 failures include nine WHITES and 46 GREENS. In terms of quarterly indicators, there are 8 WHITES and 39 GREENS. *Because some WHITE indicators remain so through several consecutive quarters, the number of unique WHITE SSU indicators is 5.*

Finally, corresponding SDP evaluations indicate 2 WHITES and 75 GREENS (or no SDP findings) for the 77 MSPI component failures. Similarly, the MSPI results indicate 2 WHITES and 75 GREENS or no SDP findings (with the proposed frontstop). However, the MSPI WHITES are for Braidwood 1 HRS and Hope Creek HPI, while the two SDP WHITES are for Millstone 2 HRS and Salem 1 EAC.

I.6   Analysis of Results

This section provides a detailed analysis of the WHITE indications, as well as near-WHITE and GREEN MSPI indications where there were a significant number of component failures.

Braidwood-1 HRS
From the 2nd quarter of 2001 through the 1st quarter of 2002, there were three failures of the diesel-driven AFW pump (DDP). Analysis indicates that given the number of demands and run-hours over the 3-year measurement interval, the expected number of failures of the DDP is approximately 1. The MSPI was GREEN for the first failure. The second failure indicated WHITE absent the use of the frontstop, but with the frontstop would be GREEN. This is consistent with the discussion in Section 5.2 whereby one failure more than baseline or expectation should not result in WHITE indication (N to N+1 issue, where N is the expected number of failures). The third failure resulted in WHITE indication regardless whether the frontstop was applied. During and shortly after this time frame, UAI contribution was significant, of the order of 3×10^{-7} to 7×10^{-7}. Thus, the WHITE MSPI indication resulted from a combination of multiple failures and large unavailability some of which accompanied those failures. It is concluded that this WHITE indication is valid, and that the MSPI performed as intended.

The SSU indicated WHITE owing to the use of a large fault exposure time of 335.8 hours as a surrogate for not directly accounting for reliability. The corresponding average system unavailability of 2.3% exceeded the generic threshold of 2.0%, thus accounting for the WHITE. It should be noted that the generic threshold of 2.0% does not account for the fact that the Braidwood design has only two AFW pumps, compared to many other PWR designs with three.

The April 29, 2002 inspection report referred to one finding of "very low safety significance (Green)" because the licensee failed to identify the cause and prevent recurrence from a previous failure.

It is concluded that for this case, the MSPI approach provides a better overall measure of system performance than the SSU. Both unavailability and unreliability contribute to the measure in the MSPI. The GREEN/WHITE threshold is exceeded in the MSPI based on consideration of plant-specific design features and performance, such as the relative risk importance of the diesel-driven AFW pump in the licensee's PRA. The frontstop behaved as intended, and because indication did not turn WHITE until three failures had occurred, the likelihood of *false positive* is low.

Hope Creek EAC
From the 2nd quarter of 2000 through the 4th quarter of 2002, there were six failures of the EDGs at Hope Creek. The plant design consists of four EDGs plus a backup gas turbine generator, and thus the relative risk-importance of an EDG failure is low. Analysis indicates that the expected number of failures (with Bayesian updating of failure rates) in the 3-year time frame for this system given the number of demands and run-hours to be about 2. The MSPI indication for this period is GREEN regardless of frontstop, with UAI of the order of -5×10^{-7} owing to better-than-baseline unavailability, and the URI varying from 1×10^{-7} to 3×10^{-7}. Sensitivity studies using the recommended generic common-cause failure (CCF) multiplier of 1.5 per Appendix F for this design configuration indicate that the inclusion of CCF would not change the color indication.

The SSU is also GREEN in this time frame. The average train unavailability reached 1.9%. The SSU does not account for the plant-specific design configuration in so far as the GREEN/WHITE threshold. The threshold is 2.5% regardless of whether there are 2 EDGs or 4 EDGs plus diverse backup power. Thus the MSPI approach is preferred in this regard. The SDP evaluations also indicated GREEN.

It is concluded that the MSPI, SSU, and SDP indication results are in congruence. Because the MSPI specifically accounts for a) unreliability and unavailability contribution to overall risk, and b) plant-specific design features including the number and relative risk-importance of the emergency and back-up power supplies, the MSPI provides a better overall measure of integrated system performance than the SSU.

Hope Creek HPI

From 3rd quarter 2000 through 2nd quarter of 2001, there were three MOV failures on the high-pressure coolant injection (HPCI) system compared to an expected number of failures much less than 1. These failures corresponded to URI of the order of 8×10^{-7}. In the 3rd quarter of 2002, about 92 hours of train unavailability along with unavailability from previous quarters was sufficient to result in UAI above 2×10^{-7}, thus placing the overall MSPI just above 1×10^{-6} (WHITE).

The SSU on the other hand peaked at about 1.7%, quite distant from the generic HPCI GREEN/WHITE threshold of 4% for boiling-water reactors (BWRs). The fact that there was no large fault exposure hours contributing to the SSU measurement explains the GREEN SSU indication. Indeed, analysis of all the WHITE SSUs for the MSPI pilot plants indicate it is always the case that large fault exposure times are the main reasons why indication is WHITE. Finally, there were no SDP evaluations associated with these MOV failures.

It is concluded that for this case, the MSPI approach provides a better overall measure of system performance than the SSU. Both unavailability and unreliability contribute to the measure in the MSPI. The GREEN/WHITE threshold is exceeded in the MSPI based on consideration of plant-specific design features and performance, such as the relative risk-importance of the HPCI MOVs. There is sufficient margin between the actual number of failures in the system (three) and expected number (a fraction of one) to conclude that there is low likelihood that this particular positive indication is a *false positive* indication. Moreover, the SSU approach failed to account for the reliability contribution to system performance — a significant deficiency in the approach. Apparently, because the MOV failures were not deemed to be the result of a licensee performance deficiency, no SDP evaluation was performed.

Millstone-2 HRS

On September 20, 2000 the turbine-driven AFW pump (TDP) failed during the normal surveillance test. The MSPI indicated that the unavailabilities of the three trains of AFW during the 3-year interval of the pilot were much better than baseline for the two motor-driven pump (MDP) trains, and about baseline for the TDP. The expected number of failures of components within the system would have been about one over the 3 years of the pilot. That is, reliability was about at-baseline on average. The net result is an MSPI value of the order of -4×10^{-7} (GREEN). This negative value is exactly as intended, allowing the risk-weighted contribution of better-than-baseline performance of the MDPs to more than offset the near-baseline performance of the TDP in the overall measure of system performance.

I-6

The SSU for this quarter was 2.7% (i.e., above the generic 2.0% GREEN/WHITE threshold). The SSU had jumped from 0.4% in the previous quarter to 2.7% owing to more than 670 hours of assumed fault exposure time associated with the TDP failure. Since the SSU does not directly account for reliability, fault exposure time is used as a surrogate. In this case, the application of the large fault exposure time in conjunction with a generic, non-risk-informed GREEN/WHITE threshold results n WHITE indication.

The discussion above describes how the MSPI for this system found the MDP trains to have better than baseline unavailability, and the TDP train to be about at baseline. The baseline unavailabilities are based on industry average unplanned unavailability from ROP data for 1999 through 2001. The planned unavailabilities are based on plant-specific values for period 1999 through 2001. A review of the baseline unavailabilities for Millstone-2 indicates that while they are higher than industry average, they are within the range of the norm (less than one standard deviation from the mean). If so, then it can not be concluded that the average AFW system unavailability for Millstone-2 is indicative of degraded performance.

With regard to component unreliability, there were 52 TDP start demands at Millstone-2 in the 3-year measurement period of the pilot. Assuming an industry-averaged failure-to-start rate of 9×10^{-3} per demand, the expected number of failures of the pump would have been 0.47. The one failure of the TDP is not inconsistent with this expectation. Assuming a constant rate of pump testing and operation, the mean-time-to-failure of the TDP would have been about 6 years. The last functional failure of the TDP was over a decade ago (1989). Thus, even accounting for the extended plant shutdown in the late 1990s, the TDP reliability performance is consistent with the industry norm.

The SDP evaluation for the TDP failure in question was also identified as WHITE. This came about from an originally assumed T/2 fault exposure time of 14 days. (The fault exposure time was later revised in the SSU to a full T or 28 days when the cause of the failure was identified.)

It is concluded that for this case, the MSPI approach provides a better overall measure of integrated system performance than the SSU. Both unavailability and unreliability contribute to the MSPI measure commensurate with the relative risk importance of these two elements. System unavailability is within the norm, and the reliability of the TDP is consistent with the industry norm as well when measured over a time period consistent with most PRA models. Given the single failure over 3 years, the likelihood that the MSPI is giving *false negative* indication is low. Rather, the inappropriate use of fault exposure time in the SSU as a surrogate for reliability is the primary reasons why the SSU appear to give *false positive* indication from an integrated system performance perspective.

Palo Verde-2 HRS
In the 4th quarter of 2000, there was a single failure of the motor-driven AFW pump to start. Because of the high risk importance of the pump, the MSPI without the frontstop was calculated to be about 3×10^{-6} (WHITE). The UAI contribution during the period of the pilot (calendar year 2002) varied between about 2×10^{-7} and 5×10^{-7}. The application of the frontstop reduced the overall MSPI to about 4×10^{-7} (GREEN) in the 4th quarter 2002, absent any inclusion of CCF effects. A sensitivity study indicated that the inclusion of CCF would not change the color indication. Should a second failure occur within the AFW system in the 3-year interval, the MSPI would very likely become WHITE (hence, a designation of near-WHITE). This is exactly as intended, based on the principle that because of the high likelihood of *false positive* indication, a single failure of a component in a 3-year interval, all other parameters at baseline, should not result in WHITE indication. In fact, the expected number of failures within the AFW system given the number of demands and run-hours in the 3-year interval is about 0.4, so an actual single failure would not be representative of degraded system performance.

The SSU for AFW was 0.5% (GREEN), with no fault exposure hours. This is distant from the 2.0% generic GREEN/WHITE threshold. There was no documented SDP evaluation.

It is concluded that for this case, the MSPI approach provides a better overall measure of system performance than the SSU. Both unavailability and unreliability contribute to the measure in the MSPI. The frontstop performed as intended by minimizing the likelihood of *false positive*. One additional failure within the system over 3 years, or additional system unavailability beyond baseline, could potentially result in WHITE indication in the MSPI. Because the MSPI (with the frontstop) and the SSU results were both GREEN, and because the one actual failure is consistent with the expected number, it is judged that the likelihood of a *false negative* on the part of the MSPI in this case is low.

<u>Palo Verde-3 HPI</u>
From 1st quarter 2000 through 4th quarter of 2001, there were two MOV failures in the high pressure safety injection system. System unavailability in this time frame was near baseline. The expected number of failures of components in the system in the 3-year interval is about 0.3. But because these valves had relatively low risk-importance (factors of 5 to 10 lower than the pumps), the MSPI remained far below the GREEN/WHITE threshold, as expected.

The SSU in the first quarter of 2000 when the first MOV failure occurred was 3.0% (WHITE), double the generic threshold of 1.5%. All of this can be attributed to an assumed T/2 fault exposure time of 984.14 hours owing to quarterly surveillance. It should be noted that because of issues associated with the use of T/2 fault exposure time as a surrogate for reliability, the use of T/2 in the SSU was discontinued in January 2002. Thus, had this MOV failure occurred some two years later, indication would have been GREEN rather than WHITE.

A supplemental inspection resulted from the PI, but there were no SDP evaluations associated with the two MOV failures.

It is concluded that for this case, the MSPI approach provides a better overall measure of system performance than the SSU. Both unavailability and unreliability are accounted for in the MSPI. Unavailability of the system was near baseline. The two MOV failures were of low risk importance, and the MSPI properly accounted for this plant-specific feature in the calculation. On the other hand, the SSU only resulted in WHITE indication because of the use of T/2 fault exposure time, a practice that was later discontinued.

<u>Salem-1 EAC</u>
During the 3rd quarter of 2002, there were four failures of the EDGs. Three of the failures were classified as failures-to-load/run (failure in less than one hour of running after successful start), and one as failure-to-run (run failure beyond one hour). Without the frontstop, the MSPI was calculated to be about 3×10^{-6} (WHITE). With the frontstop, the MSPI was around 8×10^{-7}, or near-WHITE. Inclusion of CCF would not significantly alter these results. The expected number of failures (N) is calculated to be about 2.3. Sensitivity studies indicate that the MSPI would become WHITE either on a) one additional EDG failure (of any mode) through the 2nd quarter of 2005, or b) a total of about 40 hours of additional EDG unavailability in the 3-year measurement period along with the four actual failures. In addition, the MSPI is found to be somewhat sensitive to the mode of failure of the emergency diesel generators. A sensitivity study found that had one of the failures been a failure-to-run rather than a failure-to-load/run, the MSPI would have been WHITE with or without the frontstop. Hence, the MSPI could become WHITE on four failures with additional unavailability or a different set of failure modes, or five failures at

most. This is not inconsistent with the N to N+1 principle but does illustrate the importance of classifying the failure mode. The frontstop generally had the intended effect of precluding WHITE indication on one failure more than baseline (N+1), while N+2 failures likely would indicate WHITE except for some unique combination of failure modes, whereby it would be WHITE just under N+3.

The SSU for this quarter was 1.5% (GREEN), compared to the generic GREEN/WHITE threshold of 2.5%. It should be noted that the fault exposure time of about 88 hours in the SSU was relatively low. On the other hand, the SDP evaluation associated with the September 13, 2002 failure of the turbocharger resulted in WHITE indication based on 283 hours of fault exposure time.

The results for this case amplify the differences in approach between the three measures. On one hand, the failure to properly account for unreliability resulted in underestimating the risk-impact of the EDG outages and a GREEN performance indication on the part of the SSU. On the other hand, the WHITE SDP evaluation was based on the risk impact of a single EDG failure event with a 12-day estimated fault exposure time. Multiple failures were not factored into the SDP risk evaluation.

It is judged that the near-WHITE MSPI provides a better measure of integrated system performance than the SSU. Both unreliability and unavailability are properly accounted for in the MSPI. The MSPI is more consistent with the valid SDP WHITE finding than the SSU.

San Onofre-2 SWS/CCW
From 1st quarter 2001 through 4th quarter 2002, there were six failures of the motor-driven salt water (service water) pumps. The MSPI for this system was near baseline owing to the balancing of unreliability and unavailability. However, the "backstop" for this component based on the plant-specific number of demands and run-hours, and the use of generic industry failure rates, is 7. Thus, the MSPI is a near-WHITE. One additional failure over the 3-year performance measurement period would result in WHITE indication. This is as intended, and illustrates the application of the "backstop" concept in identifying statistically significant departure of component performance from the industry norm.

There is no equivalent SSU because the cooling water support systems are not part of the current ROP. The service water cooling pump failures did not cascade sufficiently so as to cause WHITE indication. There were no SDP evaluations reported for these failures.

For this case, the MSPI approach seems to provide a valid overall measure of integrated system performance. The "backstop" in this case was nearly reached, and is appropriate for indicating statistically significant departure from the norm.

Additional SSU WHITE Indicators
In addition to the WHITE SSU indicators discussed above, there were three other cases for which the MSPIs were GREEN, and the SDP evaluations were either GREEN or there were no SDP findings:

- Surry-1 EAC: Four failures of the EDGs between 3rd quarter 2000 and 4th quarter 2002. Fault exposure time was 238 hours. SDP evaluation was GREEN.

- Surry-2 EAC: Five failures of the EDGs between 3rd quarter 2000 and 4th quarter 2001 (some of the failures are common to Surry-1). Fault exposure time was 336 hours. SDP evaluation was GREEN

These WHITE indicators were the result of the inappropriate use of fault exposure time as a surrogate for unreliability in the SSU. It is concluded that the MSPI approach provides a better overall measure of integrated system performance than the SSU. Both unavailability and unreliability contribute to the MSPI measure commensurate with the relative risk importance of these two elements.

Additional Non-GREEN Indicators Out-of-Scope of MSPI
Table I.2 identifies additional non-GREEN SSU or SDP evaluations:

- Millstone-2 HPI: Failure of the motor-driven pump in the 3rd quarter of 2000. Fault exposure time was 654 hours. SDP evaluation was GREEN.

- Prairie Island-1 and 2 SWS/CCW: In the 4th quarter of 2000, the pumps were declared inoperable because of a design condition with non-safety related power supply to the backflush system for cooling and lubrication. The SDP evaluation was WHITE.

- Surry-1 and 2 EAC: In the 2nd quarter of 2001, a degraded condition was identified on the EDGs whereby failed piston rings would have caused the diesel not to meet its mission time. The SSU was WHITE on Unit 2 owing to over 500 hours of fault exposure time. The initial SDP evaluation was WHITE, although a revised assessment of the fault exposure time may indicate YELLOW.

None of these cases were within the scope of the MSPI. The guidelines states that "conditions not capable of being discovered during normal surveillance tests" are not within scope of the MSPI, and the inspection process would be applicable. Thus, the WHITE findings for Prairie Island, and the WHITE (possibly YELLOW) results for both Surry units would remain in effect.

Summary
The MSPI, SSU, and SDP use three fundamentally different approaches. The MSPI measures statistically valid risk-informed performance of systems. It accounts for both unreliability and unavailability over a 3-year interval. Extensive research has shown that such an interval best minimizes the probability of *false positive* and *false negative* indication. The SSU directly accounts for unavailability averaged over 3 years, while indirectly attempting to address unreliability through the use of fault exposure time. The use of T/2 fault exposure time was discontinued two years ago because it contributed to arguably *false positive* indications. The SDP, on the other hand, quantifies short-term peak contributions to annual cumulative risk. This is intended to capture excessive risk contributions resulting from performance that is degraded from baseline values.

Recognizing that there are fundamental differences in approach between the MSPI, SDP, and SSU, a comparison was made of these three measures to determine whether there was overall congruence in the results. In this regard, 77 failures over 3 years as reported in the MSPI program for all pilot plants were analyzed. The quarterly indication results for the MSPI that were measurably impacted by the failures were compared to the equivalent SSU performance indication as appropriate. When an SDP finding was available for the failure in question, these results were also compared. Not surprisingly, the MSPI, SSU and SDP measures were found to be in agreement the vast majority of the time for non-risk-significant failures. However, results for non-GREEN findings and indications were mixed.

Four of the five WHITE or near-WHITE MSPI indications discussed above involved multiple failures and substantial unavailabiity contribution that, in combination, provided a high degree of confidence that system performance was at or near the point of degradation. A fifth near-WHITE was the result of multiple failures approaching the backstop, indicating pump performance bordered on statistically significant departure from industry norm.

All of the SSU non-GREEN indications were driven almost entirely by the use of fault exposure time as a surrogate for a valid reliability calculation. In one case, the indicator had turned WHITE on T/2 fault exposure time before that approach was discontinued, otherwise it would have indicated GREEN.

There were two WHITE SDP evaluations corresponding to the 77 component failures. One involved a single failure corresponding to a mean-time-to-failure based on historical performance no different than the industry norm. The other was on a system for which the SSU indicated WHITE and the MSPI indicated a near-WHITE measure. Several other non-GREEN SDP evaluations were on conditions which are out-of-scope of the MSPI, and the proposed process would default to the current inspection process and the non-GREEN findings would be applied.

Throughout all the cases studied, the MSPI appears to consistently provide a better overall measure of integrated system performance than the SSU, while minimizing both *false positive* and *false negative* likelihoods to the extent possible. Recognizing that there are differences in purpose and approach between the SDP and the MSPI, the comparisons between the two generally agreed and the differences in results are well-understood.

I.7 Reference

I.1 U.S. Nuclear Regulatory Commission,
http://www.nrc.gov/NRR/OVERSIGHT/ASSESS/index.html.

Table I.1 MSPI/SSU/SDP Comparison for MSPI Failures

Plant	System	Failure	Date	MSPI Delta CDF (1/y) (4Q2002 Data) (note 1)	MSPI Color (note 2)	SSU Unplanned Outage Time (h)	SSU Fault Exposure Time (h)	SSU Result	SSU Color	SDP Failure Mentioned in Inspection Report?	SDP Color Indicated in Inspection Report (note 3)	Comments
Braidwood 1	EAC	EDG FTS	1Q2000	-9.60E-08	Green	18.3	7.4	0.40%	Green	None		
	HPI	MOV FTO/C	3Q2001	4.39E-08	Green	35.2	0	0.60%	Green	2001010	Green	SDP result from Phase 2 analysis
	HRS	DDP FTR	2Q2001	3.84E-07	Green	0	155.9	1.50%	Green	None		MSPI using 1 FTR
		DDP FTS	4Q2001	1.33E-06	White (Green)	0	335.8	2.30%	White	2002004	Green	MSPI using 1 FTR and 1 FTS SDP result from Phase 1 analysis
		DDP FTS	1Q2002	2.28E-06	White	68.6	0	2.50%	White	2002004	Green	MSPI using 1 FTR and 2 FTS SDP result from Phase 1 analysis
Braidwood 2	EAC	EDG FTLR	1Q2002	-1.63E-07	Green	11.7	0	0.30%	Green	2002007	Green	SDP result from Phase 1 analysis
	HPI	AOV FTO/C	2Q2001	-2.00E-08	Green	0	0	0.80%	Green	None		
	HRS	DDP FTR	4Q2000	1.22E-07	Green	0	8.7	0.50%	Green	None		Event occurred during process of placing shutdown cooling in service SDP result from Phase 1 analysis
	RHR	MDP FTR	4Q2001	1.71E-07	Green	0	0	0.60%	Green	2001013	Green	
Hope Creek	EAC	EDG FTR	2Q2000	-5.23E-07	Green	11.2	336	1.40%	Green	20010127?	Green	MSPI using 1 FTR MSPI UA contribution is -4.87E-7 SDP is from Phase 1 analysis
		EDG FTR	4Q2001	-4.44E-07	Green	36.3	335.5	1.80%	Green	None		MSPI using 2 FTR MSPI UA contribution is -4.87E-7
		EDG FTR	1Q2002	-3.66E-07	Green	9.2	0	1.80%	Green	2001012	Green	MSPI using 3 FTR MSPI UA contribution is -4.87E-7 SDP result from Phase 1 analysis
		EDG FTR	3Q2002	-2.87E-07	Green	35.3	0	1.90%	Green	None		
		EDG FTS	4Q2002	-1.90E-07	Green	38.7	0	1.90%	Green	None		MSPI using 4 FTR and 2 FTS MSPI UA contribution is -4.87E-7
		EDG FTS	4Q2002	-1.90E-07	Green	40.7	0	1.90%	Green	None		MSPI using 4 FTR and 2 FTS MSPI UA contribution is -4.87E-7
	HPI	MOV FTO/C	3Q2000	-3.18E-07	Green	22.7	13	1.10%	Green	None		MSPI using 1 MOV FTO/C MSPI UA contribution is -2.81E-7
		MOV FTO/C	1Q2001	1.22E-07	Green	0	0	1.00%	Green	None		MSPI using 2 MOV FTO/C MSPI UA contribution is -2.81E-7
		MOV FTO/C	2Q2001	5.61E-07	Green	0	0	0.70%	Green	None		MSPI using 3 MOV FTO/C MSPI UA contribution is -2.81E-7
		(Note 4)	3Q2002	1.05E-06	White (Note 4)	0	0	1.70%	Green	None		MSPI using 3 MOV FTO/C MSPI UA contribution (3Q2002 data) is ≥ 2.1E-7/y
	HRS	TDP FTS	4Q2002	1.22E-07	Green	0	0	1.50%	Green	None		
	RHR	MOV FTO/C	1Q2000	1.71E-07	Green	14.2	0	1.10%	Green	2000007	Green	Event occurred while supporting HPCI and RCIC surveillances during startup SDP result from Phase 1 (?) analysis
Millstone 2	SWS/CCW	MDP FTR	1Q2001	4.32E-08	Green	N/A	N/A	N/A	N/A	2002002	Green	SDP result from Phase 1 (?) analysis
	HPI	MOV FTO/C	1Q2000	-2.65E-08	Green	0	0	0.40%	Green	None		This failure is also listed under RHR MOV FTO/C ROP UA hours listed under RHR
	HRS	TDP FTS	3Q2000	-3.91E-07	Green	30.75	677.5	2.70%	White	2000011	White	SDP result from Phase 2 and Phase 3 analysis 14-day outage assumed
	RHR	MOV FTO/C	1Q2000	3.75E-10	Green	11.06	0	0.20%	Green	None		
	SWS/CCW	AOV FTO/C	4Q2002	3.13E-07	Green	N/A	N/A	N/A	N/A	None		
Millstone 3	HPI	MDP FTR	3Q2002	-2.62E-07	Green	0.03	0	1.10%	Green	None		MSPI using 2 MDP FTR
		MDP FTR	3Q2002	-2.62E-07	Green	7.3	0	1.10%	Green	None		MSPI using 2 MDP FTR
	SWS/CCW	MDP FTS	2Q2000	1.04E-07	Green	N/A	N/A	N/A	N/A	None		

Note 1 - For system failures occurring within a single quarter, the MSPI evaluation includes all of the failures within the quarter (plus any previous failures)

Note 2 - If the proposed front stop is applied and the resulting color is different, then the color using the front stop is presented in parentheses

Note 3 - If blank, there was no identified performance deficiency requiring an SDP

Note 4 - This row was added to show that the MSPI is white using 3Q2002 data (rolling 3-year period), because of a relatively large unavailability during 3Q2002 However, the MSPI returns to green the next quarter, when the 4Q2002 data are used

Acronyms: AOV (air-operated valve), CDF (core damage frequency), DDP (diesel-driven pump), EAC (emergency ac power), EDG (emergency diesel generator), FTLR (fail to load and run for 1 hour), FTO/C (fail to open or close), FTR (fail to run), FTS (fail to start), HPI (high pressure injection system), HRS (heat removal system), MDP (motor-driven pump), MOV (motor-operated valve), MSPI (mitigating systems performance index), RHR (residual heat removal system), ROP (reactor oversight process), SDP (significance determination process), SSU (safety system unavailability process), SWS/CCW (service water system/component cooling water system), UA (unavailability)

Table I.1 MSPI/SSU/SDP Comparison for MSPI Failures (continued)

Plant	System	Failure	Date	MSPI Delta CDF (1/y) (4Q2002 Data) (note 1)	Color (note 2)	SSU Unplanned Outage Time (h)	SSU Fault Exposure Time (h)	SSU Result	SSU Color	SDP Failure Mentioned in Inspection Report?	SDP Color Indicated in Inspection Report (note 3)	Comments
Palo Verde 1	EAC	EDG FTS	2Q2002	1.10E-07	Green	27.92	15.82	0.70%	Green	None		
	HPI	MOV FTO/C	1Q2000	1.90E-09	Green	0	0	1.10%	Green	None		MSPI using 1 MOV FTO/C
		MOV FTO/C	4Q2000	2.42E-08	Green	0	0	1.10%	Green	None		MSPI using 2 MOV FTO/C
Palo Verde 2	HPI	MOV FTO/C	4Q2000	1.35E-08	Green	0	29.57	1.10%	Green	None		
	HRS	MDP FTS	4Q2000	3.02E-06	White (Green)	13.97	0	0.50%	Green	None		
Palo Verde 3	EAC	EDG FTR	2Q2000	8.89E-08	Green	0	0	0.50%	Green	2001004	Green	MSPI using 1 EDG FTR SDP results from Phase 1 (?) analysis
		EDG FTS	3Q2001	1.79E-07	Green	54.97	312.1	1.30%	Green	2001005		MSPI using 1 EDG FTR and 1 EDG FTS Failure listed in inspection report (2001005) but no mention of SDP evaluation
	HPI	MOV FTO/C	1Q2000	1.36E-09	Green	11.47	984.14	3.00%	White	None		MSPI using 1 MOV FTO/C Supplemental inspection (2000012) conducted because ROP indicator changed to white No mention of SDP evaluation
		MOV FTO/C	4Q2001	2.38E-08	Green	0	0	0.80%	Green	None		MSPI using 2 MOV FTO/C
Prairie Island 1	SWS/CCW	DDP FTS	2Q2002	1.66E-07	Green	N/A	N/A	N/A	N/A	None		MSPI using 2 DDP FTS
		DDP FTS	2Q2002	1.66E-07	Green	N/A	N/A	N/A	N/A	None		MSPI using 2 DDP FTS
		DDP FTS	3Q2002	3.52E-07	Green	N/A	N/A	N/A	N/A	None		MSPI using 3 DDP FTS
Prairie Island 2	EAC	EDG FTLR	1Q2000	-8.66E-09	Green	2.1	340.05	1.50%	Green	None		MSPI using 1 EDG FTLR R
		EDG FTS	4Q2000	1.26E-07	Green	15.17	0	1.50%	Green	None		MSPI using 1 EDG FTLR and 1 EDG FTS
		EDG FTLR	2Q2001	2.26E-07	Green	199.42	0	1.80%	Green	2001013	Green	MSPI using 2 EDG FTLR and 1 EDG FTS SDP result from Phase 2 analysis
		EDG FTS	4Q2001	3.62E-07	Green	79.87	8.78	2.30%	Green	None		MSPI using 2 EDG FTLR and 2 EDG FTS
	HRS	AOV FTO/C	3Q2001	-1.90E-08	Green	13.97	390.88	1.90%	Green	None		
Salem 1	EAC	EDG FTLR	3Q2002	2.84E-06	White (Green)	103.3	87.8	1.50%	Green	2002010	Green	MSPI using 3 EDG FTLR and 1 FTR
		EDG FTLR	3Q2002	2.84E-06	White (Green)	103.3	87.8	1.50%	Green	None		MSPI using 3 EDG FTLR and 1 FTR
		EDG FTLR	3Q2002	2.84E-06	White (Green)	103.3	87.8	1.50%	Green	None		MSPI using 3 EDG FTLR and 1 FTR
		EDG FTR	3Q2002	2.84E-06	White (Green)	103.3	87.8	1.50%	Green	2002010	White	MSPI using 3 EDG FTLR and 1 FTR SDP result in May 2003 letter from NRC to utility, referencing the results of a March 26 SERP workshop EDG 1C unavailable ≈83 hours
	HPI	MDP FTR	1Q2000	-8.34E-09	Green	41.1	0	0.60%	Green	None		
	SWS/CCW	MDP FTR	2Q2000	-1.14E-07	Green	N/A	N/A	N/A	N/A	None		
Salem 2	HPI	MDP FTS	1Q2000	4.20E-08	Green	11	0	0.50%	Green	None		
	SWS/CCW	MOV FTO/C	1Q2001	1.44E-07	Green	N/A	N/A	N/A	N/A	2000011		Inspection report 2000011 discusses failure of similar valve (21SW127) in Unit 1 on 1/24/01 For that event, the other HX was already unavailable, so both CCW HXs were unavailable For this simultaneous outage, the SDP result of green was from a Phase 3 analysis (Phase 2 workbooks for Salem not available at the time) The same inspection report describes the Unit 2 failure of 22SW127 on 1/4/01, but does not mention any SDP evaluation

Note 1 - For system failures occurring within a single quarter, the MSPI evaluation includes all of the failures within the quarter (plus any previous failures)

Note 2 - If the proposed front stop is applied and the resulting color is different, then the color using the front stop is presented in parentheses

Note 3 - If blank, there was no identified performance deficiency requiring an SDP

Acronyms: AOV (air-operated valve), CDF (core damage frequency), DDP (diesel-driven pump), EAC (emergency ac power), EDG (emergency diesel generator), FTLR (fail to load and run for 1 hour), FTO/C (fail to open or close), FTR (fail to run), FTS (fail to start), HPI (high pressure injection system), HRS (heat removal system), MDP (motor-driven pump), MOV (motor-operated valve), MSPI (mitigating systems performance index), RHR (residual heat removal system), ROP (reactor oversight process), SDP (significance determination process), SSU (safety system unavailability), SWS/CCW (service water system/component cooling water system), UA (unavailability)

Table I.1 MSPI/SSU/SDP Comparison for MSPI Failures (continued)

Plant	System	Failure	Date	MSPI Delta CDF (1/y) (4Q2002 Data) (note 1)	MSPI Color (note 2)	SSU Unplanned Outage Time (h)	SSU Fault Exposure Time (h)	SSU Result	SSU Color	SDP Failure Mentioned in Inspection Report?	SDP Color Indicated in Inspection Report (note 3)	Comments
San Onofre 2	HPI	MDP FTS	3Q2000	-2.05E-08	Green	6.3	0	0.80%	Green	None		
	SWS/CCW	MDP FTR	1Q2001	-2.02E-07	Green	N/A	N/A	N/A	N/A	None		MSPI using 1 SWS MDP FTR
		MDP FTR	4Q2001	-1.64E-07	Green	N/A	N/A	N/A	N/A	None		MSPI using 3 SWS MDP FTR
		MDP FTR	4Q2001	-1.64E-07	Green	N/A	N/A	N/A	N/A	None		MSPI using 3 SWS MDP FTR
		MDP FTR	1Q2002	-1.46E-07	Green	N/A	N/A	N/A	N/A	None		MSPI using 4 SWS MDP FTR
		MDP FTR	4Q2002	-9.53E-08	Green	N/A	N/A	N/A	N/A	None		MSPI using 6 SWS MDP FTR
		MDP FTR	4Q2002	-9.53E-08	Green	N/A	N/A	N/A	N/A	None		MSPI using 6 SWS MDP FTR
San Onofre 3	SWS/CCW	MDP FTR	3Q2001	-4.81E-07	Green	N/A	N/A	N/A	N/A	None		MSPI using 2 SWS MDP FTR
		MDP FTR	3Q2001	-4.81E-07	Green	N/A	N/A	N/A	N/A	None		MSPI using 2 SWS MDP FTR
Surry 1	EAC	EDG FTLR	3Q2000	2.00E-07	Green	83.68	0	1.50%	Green	None		MSPI using 1 EDG FTLR and 1 EDG FTR
		EDG FTR	3Q2000	2.00E-07	Green	83.68	0	1.50%	Green	None		MSPI using 1 EDG FTLR and 1 EDG FTR
		EDG FTS	4Q2001	2.96E-07	Green	0.5	237.77	2.70%	White	2001007	Green	MSPI using 1 EDG FTLR, 1 EDG FTR, and 1 EDG FTS SDP result from Phase 1 (?) analysis
		EDG FTS	4Q2002	3.91E-07	Green	85.35	0	3.20%	White	None		MSPI using 1 EDG FTLR, 1 EDG FTR, and 2 EDG FTS ROP/SSU for Unit 2 (EDG shared by both units) is 2.9% and white
Surry 2	EAC	EDG FTS	3Q2000	4.58E-08	Green	18.94	0	1.80%	Green	None		MSPI using 1 EDG FTS and 1 EDG FTLR
		EDG FTLR	3Q2000	4.58E-08	Green	18.94	0	1.80%	Green	None		MSPI using 1 EDG FTS and 1 EDG FTLR
		EDG FTLR	3Q2001	1.31E-07	Green	22.32	336.03	3.10%	White	2002008	Green	MSPI using 1 EDG FTS and 2 EDG FTLR SDP result from Phase 1 (?) analysis
		EDG FTLR	4Q2001	4.00E-07	Green	133.15	0	3.20%	White	None		MSPI using 1 EDG FTS and 4 EDG FTLR
		EDG FTLR	4Q2001	4.00E-07	Green	133.15	0	3.20%	White	None		MSPI using 1 EDG FTS and 4 EDG FTLR
Surry 1/2	SWS/CCW	DDP FTR	2Q2000	<1.97E-07	Green	N/A	N/A	N/A	N/A	None		Surry 1 MSPI using 1 DDP FTR
		DDP FTS	3Q2000	<1.97E-07	Green	N/A	N/A	N/A	N/A	None		Surry 1 MSPI using 1 DDP FTR and 1 DDP FTS
		MOV FTO/C	2Q2001	<1.97E-07	Green	N/A	N/A	N/A	N/A	None		Surry 1 MSPI using 1 DDP FTR, 1 DDP FTS, and 1 MOV FTO/C
		DDP FTS	2Q2002	<1.97E-07	Green	N/A	N/A	N/A	N/A	None		Surry 1 MSPI using 1 DDP FTR, 2 DDP FTS, and 1 MOV FTO/C
		MOV FTO/C	4Q2002	1.97E-07	Green	N/A	N/A	N/A	N/A	None		Surry 1 MSPI using 1 DDP FTR, 2 DDP FTS, and 3 MOV FTO/C
		MOV FTO/C	4Q2002	1.97E-07	Green	N/A	N/A	N/A	N/A	None		Surry 1 MSPI using 1 DDP FTR, 2 DDP FTS, and 3 MOV FTO/C

Note 1 - For system failures occurring within a single quarter, the MSPI evaluation includes all of the failures within the quarter (plus any previous failures)

Note 2 - If the proposed front stop is applied and the resulting color is different, then the color using the front stop is presented in parentheses

Note 3 - If blank, there was no identified performance deficiency requiring an SDP

Acronyms: AOV (air-operated valve), CDF (core damage frequency), DDP (diesel-driven pump), EAC (emergency ac power), EDG (emergency diesel generator), FTLR (fail to load and run for 1 hour), FTO/C (fail to open or close), FTR (fail to run), FTS (fail to start), HPI (high pressure injection system), HRS (heat removal system), MDP (motor-driven pump), MOV (motor-operated valve), MSPI (mitigating systems performance index), RHR (residual heat removal system), ROP (reactor oversight process), SDP (significance determination process), SSU (safety system unavailability), SWS/CCW (service water system/component cooling water system), UA (unavailability)

Table I.2 Additional SSU and SDP Whites not Listed in Table I.1

Plant	System	Failure	Date	MSPI		SSU				SDP		Comments
				Delta CDF (1/y) (4Q2002 Data) (note 1)	Color (note 2)	Unplanned Outage Time (h)	Fault Exposure Time (h)	Result	Color	Failure Mentioned in Inspection Report?	SDP Color Indicated in Inspection Report (note 3)	
Millstone 2	HPI	MDP FTR	3Q2000	N/A	N/A	0	654 2	3 10%	White	2000011	Green	This condition event was discovered during periodic testing Pump operation beyond 4 hours was determined to be questionable Not an MSPI failure SDP modeled recovery by placing a spare pump into service
Prairie Island 1	SWS/CCW	MDP FTR	4Q2000	N/A	N/A	N/A	N/A	N/A	N/A	2000013	White	No safety-related electrical power to backwash system Not an MSPI failure
Prairie Island 2	SWS/CCW	MDP FTR	4Q2000	N/A	N/A	N/A	N/A	N/A	N/A	2000013	White	No safety-related electrical power to backwash system Not an MSPI failure
Surry 1	EAC	EDG FTR	2Q2001	N/A	N/A	131 35	192 03	2 10%	Green	2001006	White (Yellow)	Degraded condition identified during disassembly of EDG Not an MSPI failure SDP initially yellow but later changed to white Recent SDP Phase 3 analysis indicates yellow with a longer fault exposure time
Surry 2	EAC	EDG FTR	2Q2001/ 3Q2001	N/A	N/A	131 35 + 22 32	192 03 + 336 03	3 10%	White	2001006	White (Yellow)	Degraded condition identified during disassembly of EDG Not an MSPI failure SDP initially yellow but later changed to white Recent SDP Phase 3 analysis indicates yellow with a longer fault exposure time

Note 1 - For system failures occurring within a single quarter, the MSPI evaluation includes all of the failures within the quarter (plus any previous failures)

Note 2 - If the proposed front stop is applied and the resulting color is different, then the color using the front stop is presented in parentheses

Note 3 - If blank, there was no identified performance deficiency requiring an SDP

Acronyms: AOV (air-operated valve), CDF (core damage frequency), DDP (diesel-driven pump), EAC (emergency ac power), EDG (emergency diesel generator), FTLR (fail to load and run for 1 hour), FTO/C (fail to open or close), FTR (fail to run), FTS (fail to start), HPI (high pressure injection system), HRS (heat removal system), MDP (motor-driven pump), MOV (motor-operated valve), MSPI (mitigating systems performance index), RHR (residual heat removal system), ROP (reactor oversight process), SDP (significance determination process), SSU (safety system unavailability), SWS/CCW (service water system/component cooling water system), UA (unavailability)

APPENDIX J. TECHNICAL BASIS FOR USING THE CONSTRAINED NON-INFORMATIVE PRIOR

Appendix J
Technical Basis for Using the Constrained Non-Informative Prior

J.1 Introduction

Assessment of current performance is very different from assessment of long-term average performance. But most PRA-related data analysis is concerned primarily with long-term average performance; it typically reflects an assumption that the parameters being estimated are essentially static. Much of it regards data from different sources as being representative of a homogeneous population, or at least considers the mean values of performance parameters extracted from these sources to be the quantities of interest, and the right quantities to use in PRA. Even "population variability" methods, while recognizing that performance varies from one member of the population to another, are typically aimed at extracting long-time averages of performance parameters. Such quantities are long-time averages over different *performance states.*

The problem of determining whether *current* performance deviates from historical norms, based on sparse current data, is more difficult than estimating a long-term average. In many problems of interest, although a significant body of historical evidence is available, current performance information is too sparse to be the sole basis for an assessment of how well the system is currently performing. Therefore, t is desirable to apply current data within a Bayesian framework, making use of a broader body of evidence related to performance. The "constrained non-informative prior" (CNIP) (Ref. J.1) does this in the MSPI. This appendix summarizes why the CNIP works as well as it does and the basis for its selection, indicates where there is room for improvement, and suggests possible future directions.

J.2 Prior Distributions Evaluated

All approaches discussed within this section are Bayesian, in that they formulate a prior distribution on performance parameters,[1] update these distributions with current data to derive posterior distributions of current values of performance parameters, and then use information from the posteriors in a decision rule. NUREG-1753 (Ref. J.2) studied several ways of using prior information to estimate current performance:

• Update the "Industry" Prior	The industry prior reflects variability across the industry of the long-term average value.
• Update the Constrained Non-Informative Prior (CNIP)	Mean of the prior distribution is the industry mean. Other characteristics of the prior are determined by the requirement to be "non-informative." This prior is updated with current failure and demand information.
• Maximum-Likelihood Estimate (MLE)	Makes no use of historical information; derives an estimate entirely from current failure and demand information. This is non-informative in an intuitive sense, but true "non-informative" priors actually need to have more complex mathematical properties.

---

[1] Even the MLE can be thought of in this way. For the demand failure probability case, the MLE is like having previously observed zero failures in zero demands, which we "update" with current data by adding current failures to the (zero) numerator and current demands to the (zero) denominator. For this reason, NUREG-1753 actually refers to the MLE as being based on a "zero" prior.

The assessment process used in NUREG-1753 was the following:

(1) Begin with a "baseline" value of unreliability, corresponding to industry average behavior. Build this value into a prior distribution on unreliability.

(2) Update this prior with current performance information: for example, the number of demands n and number of failures x observed in a particular component group within a particular assessment period (time window).

(3) Take the mean of the posterior distribution as the estimate of current unreliability. Subtract from this the baseline value, in order to obtain an estimate of the change in unreliability.

(4) Multiply the change in unreliability by the associated Birnbaum importance to obtain an estimate of the change in the applicable risk metric ("core damage frequency," in the case of NUREG-1753).

(5) Compare the change in the risk metric with decision thresholds to determine the appropriate programmatic response.

Unless the number of observed failures x is fairly large, the scatter in x is significant compared to x itself. For many cases of practical interest in this program, x is not large, even when performance is degraded. Therefore, using a maximum-likelihood estimate of current unreliability (x/n, dividing observed failures by observed demands) gives rise to a noisy signal. One implication of this is a high probability of a false indication of declining performance, which wastes resources in regulatory and licensee response, and creates issues of false perceptions. On the other hand, using a prior that is narrowly focused on the baseline estimate strongly biases the posterior towards that baseline; if performance changes significantly, much data will be required to shift the distribution to the right area.

NUREG-1753 compared the behavior of the CNIP with the MLE and with the "industry prior," with respect to their respective efficacies in the above decision rule. The behavior of each alternative was investigated in specific postulated scenarios. Given baseline unreliability performance, the conditional probability of falsely assessing degraded performance was determined, and similarly for the probability of falsely assessing "good" performance given that performance was actually degraded. In NUREG-1753, the CNIP was found to be the best of the alternatives considered at that time. The MLE has a false-positive problem: it uses the number of failures directly, and as indicated above, this is a noisy signal. The "industry prior" has the opposite problem: it gives less prior density to large excursions, creating a false negative potential. The CNIP falls between these extremes and provides the best combination of minimizing both *false positive* and *false negative*.

Although the CNIP is an improvement over the other alternatives, using the CNIP in the above process still yields a significant false-indication probability in many cases of practical interest. Therefore, in some cases, a small number of failures can trigger a regulatory response, even if the failures occurred within the observation window by coincidence, rather than because of declining performance. In other cases, a low value of the CNIP density for high failure probability requires the accumulation of a significant number of failures before the posterior density becomes significant in that region. Because of the form of the CNIP, if the baseline failure probability is a very small number, the CNIP accords a very low prior probability to significantly degraded performance, and it takes a certain amount of data to overcome this. Notwithstanding these shortcomings of the CNIP, the results of the pilot program indicate that the CNIP generally provides reasonable overall results.

J.3   Research on Advanced Prior Distributions

Research into the ideal approach to address the issues discussed above is ongoing. A purist Bayesian approach would integrate all available information into a prior that reflected a considered assessment of how likely performance is to be degraded, how bad performance is when it is degraded, and how good performance is when it is good. Preliminary work has been done to explore the behavior of decision rules based on such a prior. One such formulation, a "mixture prior" (Ref. J.3), has shown real promise in reducing the potential for false indications. For small numbers of failures, the posterior distribution from updating the mixture prior is not much different from the prior, so the false positive failure probability is reduced. For larger numbers of failures, the posterior distribution switches over to reflect a significant probability of degraded performance. These characteristics are highly desirable for this performance assessment application.

However, these benefits of the mixture prior come at a certain price in complexity and in data required to support development of the prior. The CNIP is determined by only one parameter: its mean value. Given the mean, other parameters of the CNIP are determined from the requirement that the function be non-informative in a certain mathematical sense. In the MSPI process, for each component type, the mean of the associated CNIP is currently taken to be the long-term industry-average behavior of that component type. For many component types, estimates of this value can be developed. More flexible priors, such as the mixture prior, involve more parameters. Thus, although the mixture prior itself is not difficult to work with analytically, assessment of these additional parameters for applicable component types would need some work. Consensus would need to be developed regarding the characteristics of "good" and "degraded" performance, including prior probabilities of these conditions. Most available data have not been collected or analyzed with this kind of application in mind. These sorts of reasons are generically why non-informative priors are discussed in the first place.

J.4   Conclusions Regarding the Use of the CNIP

The technical basis for using a Bayesian framework for performance parameters is well-founded. NUREG-1753 identified that of the practical options that were considered, the CNIP displayed the best characteristics from the perspective of minimizing both *false positive* and *false negative* indication. The CNIP can be practically implemented because of the simple algebraic formulation resulting from the update process. It is recognized that the CNIP is not perfect, but the results of the pilot program as documented in this report indicate that the overall results are reasonable. Moreover, to address possible concerns with residual issues of *false positive* and *false negative* that could arise from the mathematical formulation of the MSPI including the CNIP, the concepts of "frontstop" and "backstop" have been proposed. These effectively constrain the minimum and maximum number of failures of components within a system that result in WHITE performance to ensure reasonableness of results. At present, the CNIP is programmatically the best available alternative, while research into improved methods continues.

J.5   References

J.1   C.L. Atwood, "Constrained Noninformative Priors in Risk Assessment," *Reliability Engineering and System Safety*, Vol. 53, No. 1, pp 37–46, 1996.

J.2   H.G. Hamzehee, et al., U.S. Nuclear Regulatory Commission (NRC). NUREG-1753, "Risk-Based Performance Indicators: Results of Phase 1 Development." NRC: Washington, DC. April 2002.

J.3   C.L. Atwood and R.W. Youngblood, "Application of Mixture Priors to Assessment of Performance," *Probabilistic Safety Assessment and Management (PSAM 7-ESREL '04)*, pp. 444-449, edited by Cornelia Spitzer, Ulrich Schmocker, Vinh Dang (Springer), 2004.

APPENDIX K. SENSITIVITY STUDIES

Appendix K
Sensitivity Studies

Three sensitivity studies were performed to determine potential impacts on the Mitigating System Performance Index (MSPI) if (1) a valve were missing from the system scope, (2) demands or run hours were overestimated, or (3) a component failure were missed. The reports on these are presented in the following three sections.

K.1   Summary of Missing Valve Impact on MSPI ΔCDF

K.1.1 Methodology

A sensitivity study was performed to estimate the potential impact on the Mitigating Systems Performance Index if a valve were missed in the system scoping effort. This study was performed by identifying all motor-operated valves (MOVs) and air-operated valves (AOVs) reported as in-scope by the 20 pilot plants. For those valves with non-zero Birnbaum importance measures, a database was generated containing the plant name, system, type of valve, Birnbaum importance, and component group number of demands (over the 3-year period ending 12/31/2002). (Of the 509 valves with non-zero Birnbaums, 426 are MOVs and 83 are AOVs.) Then for each valve, the associated MSPI delta core damage frequency (ΔCDF) was simulated. The MSPI calculation is as follows:

$$\text{MSPI}_{\text{missing valve}} \quad = \quad \text{Birnbaum}_{\text{missing valve}} (\text{UR}_{\text{current}} - \text{UR}_{\text{baseline}})$$

where $\text{UR}_{\text{current}}$ $=$ $(a + n)/(a + b + d)$

$\text{UR}_{\text{baseline}}$ $=$ baseline unreliability for the missing valve

a and b $=$ constrained noninformative prior beta distribution parameters

n $=$ simulated number of failures for component type

d $=$ number of demands for component type.

In the simulations, a, b, d, and $\text{UR}_{\text{baseline}}$ are fixed for each type of valve. The MOVs have a mean baseline failure to open or close probability of 7.0×10^{-4}, while the AOVs have a mean baseline failure probability of 1.0×10^{-3}. These baseline failure probabilities also determine the a and b parameters. Finally, the number of demands for the component type is fixed by the type of valve and the system, but ranges from 7 to 1,241 over the 3-year period (ending with the 4th quarter of 2002).

The simulated number of failures, n, was determined by sampling from a beta distribution for the baseline failure rate and then sampling from a binomial distribution using the number of demands and the sample baseline failure rate. (See Appendix L for more details concerning the simulation work.) The MSPI for each valve was simulated using 1,000 samples and a Latin hypercube sampling routine.

K.1.2 Results

For each valve MSPI simulation, the 95% ΔCDF was determined. These were then ordered from highest to lowest. These results are plotted in Figure K.1. The highest 95% ΔCDF (from all of the 509 valves) is 4.2×10^{-8}/year. Therefore, if a valve were missed in the scoping effort, a reasonable upper bound estimate for the impact on the system MSPI is 4.2×10^{-8}/year.

K.2   Impact of Overestimating Demands and Run Hours on the MSPI

K.2.1 Methodology

A sensitivity study was performed to estimate the potential impact on the MSPI if the demands or run hours for a component type failure mode were overestimated. The approach used to accomplish this increased the 4th quarter 2002 demand or run-hour data by 20% and 100%, and then simulations were run for both cases.

The simulations were performed in a fashion similar to that for the missing valve study. That is, the simulated number of failures, n, was determined by sampling from a beta distribution for the baseline failure rate and then sampling from a binomial or Poisson distribution using the number of demands or run hours and the sample baseline failure rate. The MSPI was simulated using 1,000 samples and a Latin hypercube sampling routine.

K.2.2 Results

For each simulation of the overestimation of the demands on the MSPI, the 95% ΔCDFs were determined. These were then ordered from highest to lowest. The results are plotted in Figure K.2. The highest 95% ΔCDFs from all 408 component-type failure modes were 2.3×10^{-7}/year and 7.5×10^{-7}/year, for the 20% and 100% overestimates, respectively. Of the eight (out of 408) points in the 20% overestimate case with ΔMSPI > 1.0×10^{-7}/year, six are from emergency diesel generators (EDGs) and heat removal system (HRS) diesel-driven pumps (DDPs). The 26 points in the 100% overestimate case with ΔMSPI > 1.0×10^{-7}/year include the same eight component-type failure modes from the 20% overestimate case. Also included in the 100% overestimate case are six motor-driven standby pumps (MDP-Stdby) and four turbine-driven auxiliary feedwater pumps from HRS, plus two DDPs, two MOVs, one AOV, and one MDP from service water systems.

K.3   Missing Failure Effect on MSPI

K.3.1 Methodology

A sensitivity study was performed to estimate the potential impact on the MSPI if a component failure were missed. To accomplish this the MSPI was compared before and after adding a failure to 3 years of plant data ending with the 4th quarter of 2002, for each system component-type failure mode in each of the 20 pilot plants. This comparison was also made for cases including common-cause failure (CCF) (see Appendix F), and frontstop or risk cap (RC) (see Appendix D), both individually and together. The change in MSPI CDF was then ranked in descending order for each case.

The baseline results against which comparisons were made were the plant MSPI results for the 4th quarter of 2002. (See Appendix A for a summary of these results.) The results were summarized for all unique plant system component-type failure mode combinations. There were 408 such unique combinations in the pilot program. Also, from these, the results for the 98 unique valve combinations (MOV or AOV) were summarized separately for comparison to the missing valve sensitivity study results.

K.3.2 Results

The results are plotted in Figure K.3. The 133 pilot plant ΔMSPIs $> 1 \times 10^{-6}$/year represent 44 unique plant system component-type failure mode combinations. HRS MDP-Standbys and DDPs plus EDGs account for 22 of these combinations. The highest ΔCDFs were 3.9×10^{-6}/year, 2.7×10^{-6}/year, 7.6×10^{-6}/year and 3.3×10^{-6}/year for the cases without CCF or RC, with RC, with CCF, and with RC and CCF, respectively. The last result with RC and CCF of 3.3×10^{-6}/year is the worst-case impact on the proposed MSPI for missing a failure in the pilot program.

A subset of these results for just the valves (AOVs and MOVs) is plotted in Figure K.4. By comparison to Figure K.1, it can be seen that missing a valve failure has potentially more impact on the MSPI than missing a valve in the system scoping.

Figures K.3 and K.4 indicate that including CCF makes the MSP more sensitive (both to errors and to true failures), while including the risk cap makes the MSP less sensitive (both to errors and to true failures). There is a necessary tradeoff between being sensitive to the data, so that real degradations can be recognized (desirable), and being sensitive to errors, so that data errors have a noticeable effect (undesirable). With CCF one gets more sensitivity and with the risk cap one gets less sensitivity. In each case, there are advantages and disadvantages.

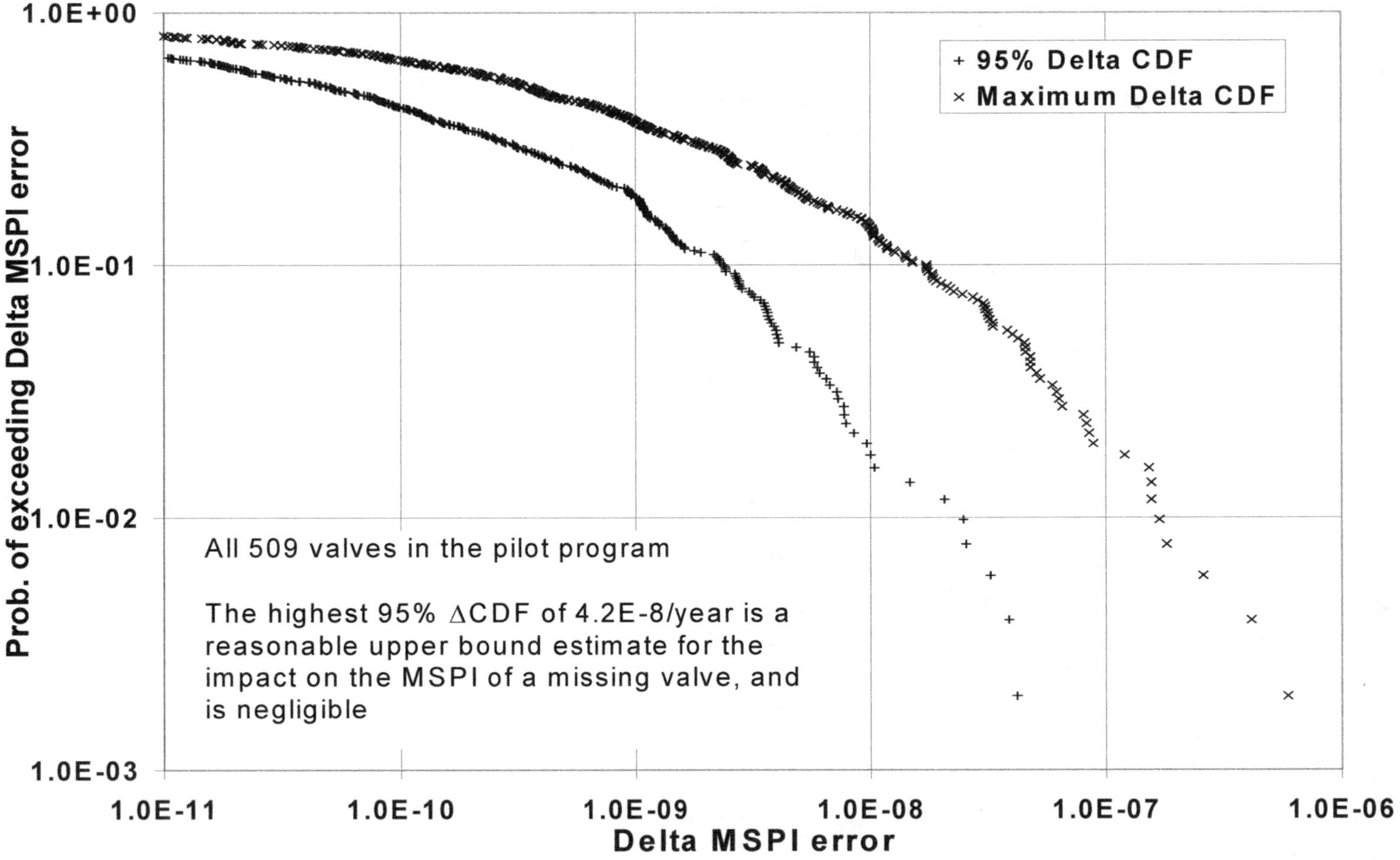

Figure K.1 Impact on the MSPI of Missing a Valve during the System Scoping

K-4

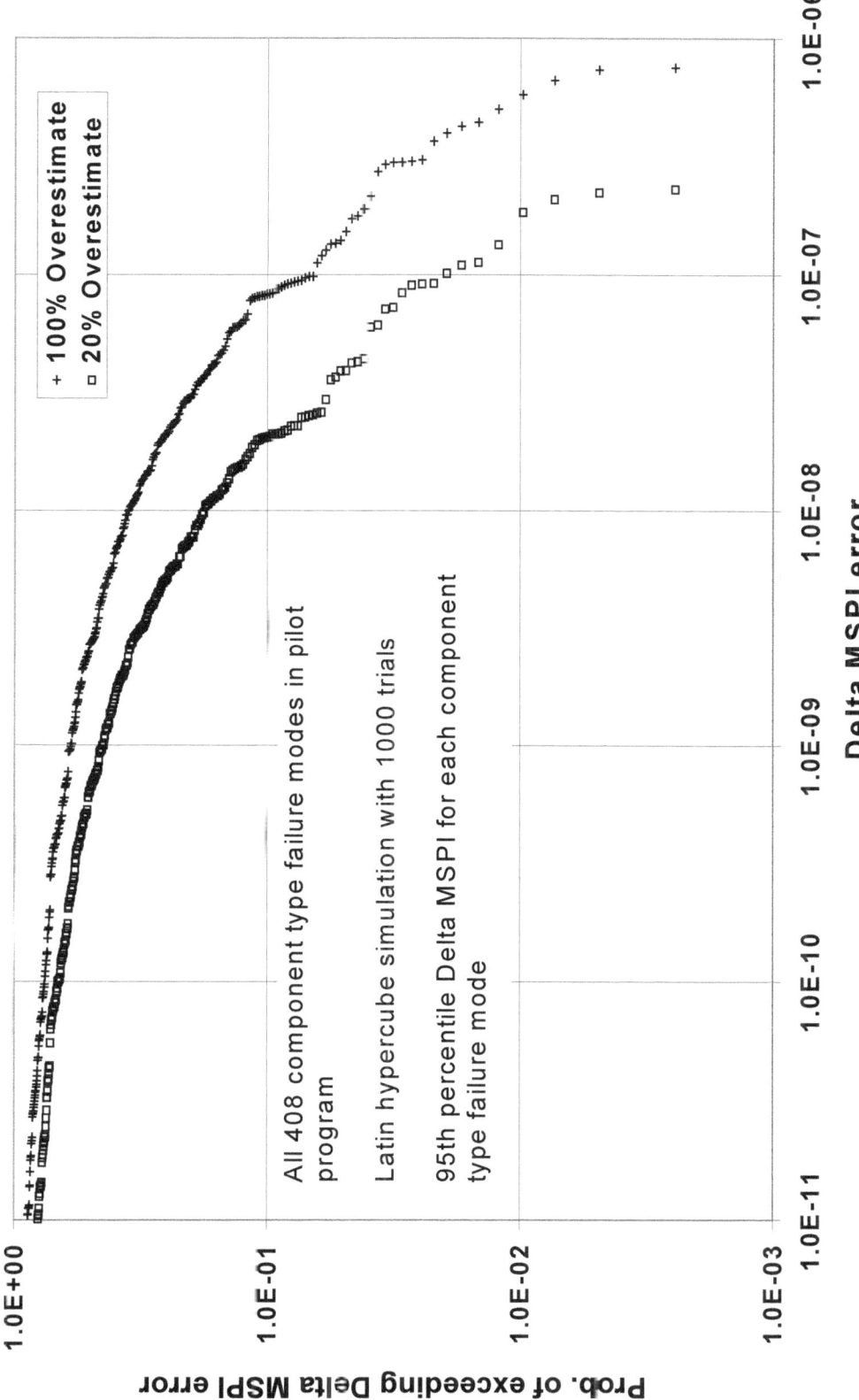

Figure K.2 Impact on the MSPI of Overestimating Component-Type Demands or Run-Hours

K-5

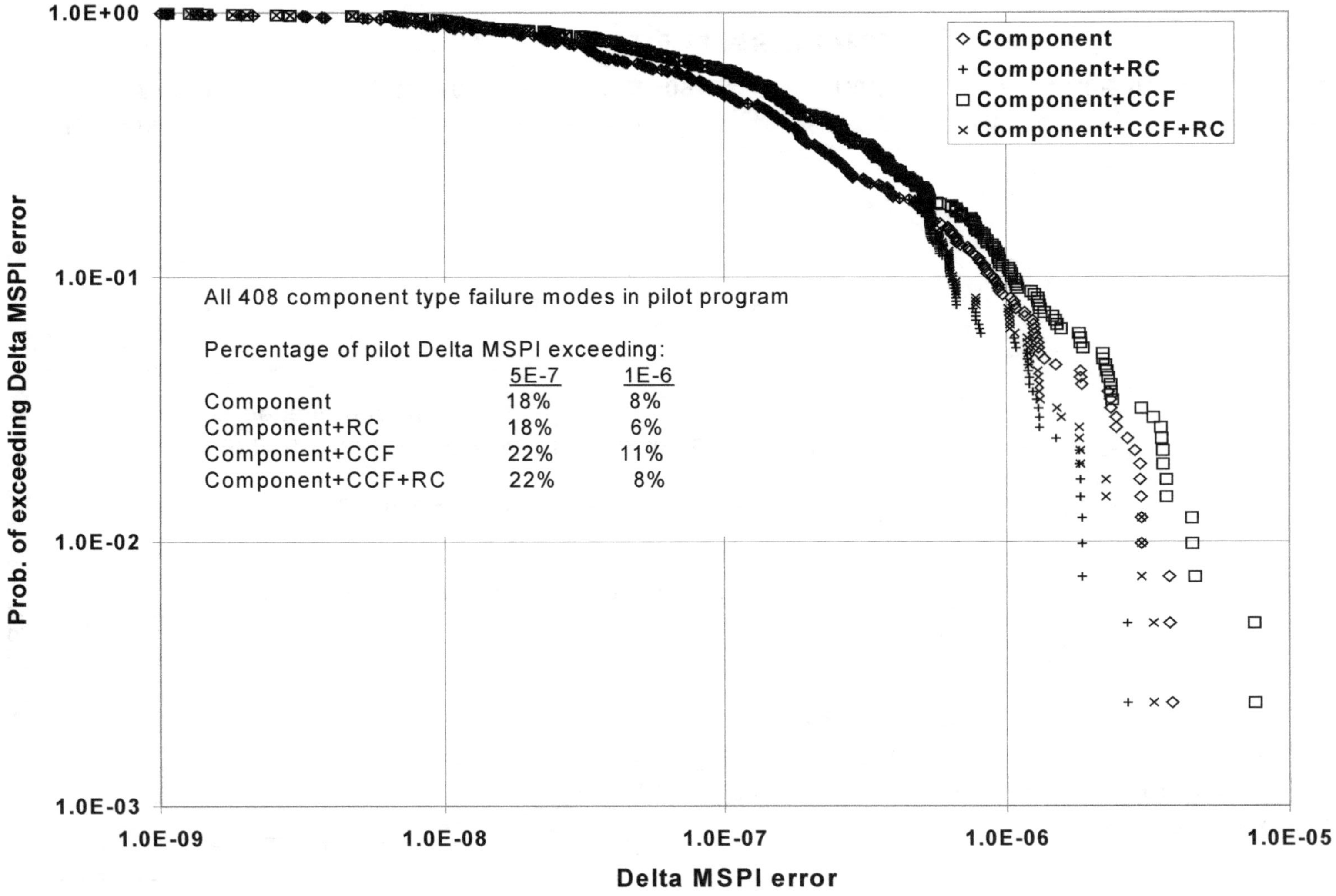

Prob. of exceeding Delta MSPI error

Delta MSPI error

Legend:
- ◇ Component
- + Component+RC
- □ Component+CCF
- × Component+CCF+RC

All 408 component type failure modes in pilot program

Percentage of pilot Delta MSPI exceeding:

	5E-7	1E-6
Component	18%	8%
Component+RC	18%	6%
Component+CCF	22%	11%
Component+CCF+RC	22%	8%

Figure K.3 Impact on the MSPI from Missing a Failure

K-6

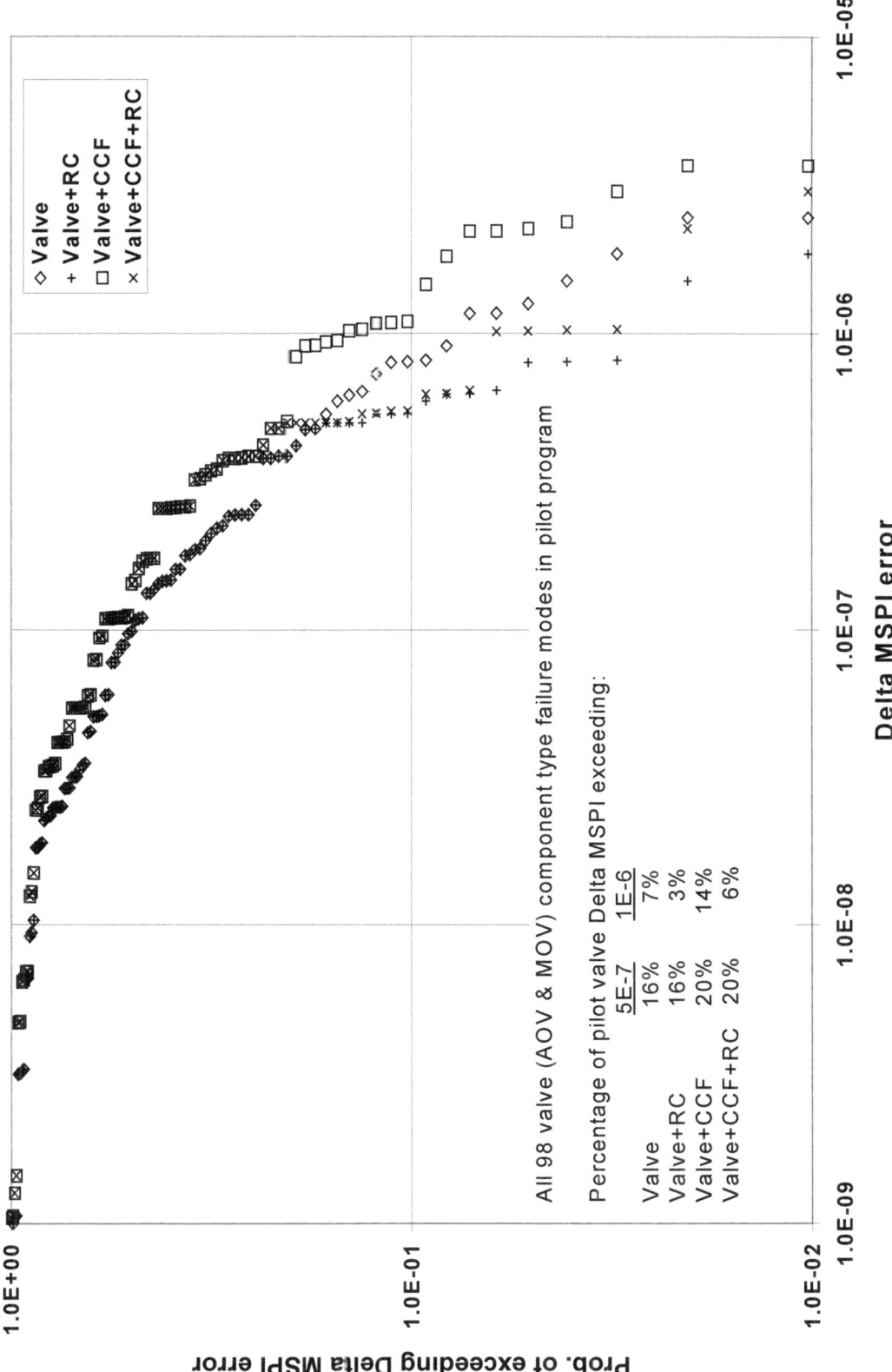

Figure K.4 Impact on MSPI from Missing a Valve Failure

K-7

APPENDIX L. SUMMARY OF MSPI SIMULATION EFFORT

Appendix L
Summary of MSPI Simulation Effort

L.1 Introduction

The Mitigating Systems Performance Index (MSPI) is a measure of approximate change in core damage frequency (CDF) resulting from changes in mitigating system component unreliability performance and train unavailability. The MSPI is evaluated for six mitigating systems at each pilot plant, with cooling water support systems combined into one indicator. For each mitigating system, the MSPI equation is the following:

$$MSPI = \left(CDF_P\right)\left(\sum \frac{FV_P}{UR_P}\right)(UR_C - UR_B) + \left(CDF_P\right)\left(\sum \frac{FV_P}{UA_P}\right)(UA_C - UA_B) \quad \text{(Eq. L.1)}$$

where
MSPI	=	*ΔCDF for the system (from changes in component UR and train UA)*
CDF_P	=	*internal events core damage frequency per calendar year (from plant PRA)*
FV_P	=	*Fussell-Vesely importance measure of the component or train (from plant PRA)*
UR_P	=	*component unreliability (from plant PRA)*
UR_C	=	*current component unreliability (Bayesian update using data from most recent 3 years)*
UR_B	=	*baseline component unreliability*
UA_P	=	*train unavailability (from plant PRA)*
UA_C	=	*current train unavailability (data from most recent 3 years)*
UA_B	=	*baseline train unavailability.*

The first summation in Equation L.1 is over all monitored components within the system, while the second summation is over all trains within the system.

Appendix A presents the actual MSPI results for the five indicators at each of the 20 pilot plants. MSPI calculations using the plant risk model inputs (CDF_P, FV_P, UR_P, and UA_P) and 3 years of data ending with the fourth quarter of 2002 indicate 3 WHITE MSPIs out of 100 total. If the proposed frontstop discussed in Appendix D is applied to the results, then there is only 1 WHITE out of 100. Application of the proposed backstop discussed in Appendix E does not change the 1 WHITE out of 100.

The actual MSPI results presented in Appendix A are the result of the 77 component failures identified in monitored components and the train unavailabilities over the 3-year period from 2000 through 2002. Another 3-year period of data collection would result in a different set of failures and resultant MSPI colors. To better understand the performance characteristics of the MSPI, simulations of future performance were performed. The simulations were performed to estimate the fraction of non-GREEN MSPIs expected, the reduction in non-GREEN MSPIs if the proposed frontstop is applied, and the additional non-GREEN MSPIs expected if the proposed backstop is applied.

The simulations discussed in this appendix were performed assuming the future performance of monitored components would follow the baseline performance observed over 1999 – 2001 but with fluctuations allowed within the baseline performance distributions. If future component performance were to significantly improve or degrade compared with baseline assumptions, the simulation results presented in this appendix would no longer be valid.

L.2 Simulation Methodology

To simulate the MSPI in Equation L.1, all values are kept constant (and at baseline conditions) except for UR_c and UA_c. The expressions for UR_c are the following:

$$UR_C = \frac{(a+N)}{(a+b+D)} \quad \text{for demand failure modes} \qquad \text{(Eq. L.2)}$$

$$UR_C = \frac{(a+N)}{(b+T)} \quad \text{for failures to run} \qquad \text{(Eq. L.3)}$$

where a, b = parameters of the baseline beta (for demand failure modes)
                              or gamma (for failures to run) distribution
         N = simulated number of failures over the 3-year time period
         D = number of demands over the 3-year period
         T = number of run hours over the 3-year period.

In these expressions, a and b remain constant and represent the constrained non-informative distributions of the Year 2000 component failure rate baselines discussed in Appendix C. Also, D and T remain constant and were obtained from the 3-year data set ending in the 4th quarter of 2002. Uncertainty in D and T was not modeled in the simulations because these values probably do not vary by more than 20 percent over a 3-year period. However, N is allowed to vary with each sample of the simulation.

For demand failure modes, N is assumed to vary randomly, following a binomial distribution with parameters D and the baseline mean failure probability. Also for the simulation of N, uncertainty in the baseline failure probability was assumed to follow a beta distribution with parameters a and b (from Appendix C). This simulation therefore includes both aleatory uncertainty in N (the randomness of the event count over 3 years) and epistemic uncertainty (lack of perfect knowledge of the value of the baseline failure probability).

For failures to run, N is assumed to vary randomly, following a Poisson distribution with the single parameter (T)*(baseline mean failure rate). Uncertainty in the baseline failure rate was assumed to follow a gamma distribution with parameters a and b (from Appendix C).

The expression for UA_C is the following:

$$UA_C = UA_B + \frac{N * 36h}{Criticalhours} \qquad \text{(Eq. L.4)}$$

where 36h   = repair time assumed for each failure
         Criticalhours = reactor critical hours over the 3-year period.

For simulation of UA_C, all values are kept constant except for N.

To simulate the system MSPI for a plant, sampling is performed at the component-type and failure mode level, rather than at the individual component and failure mode level. This is because the data for similar components within a component type (failures and demands or run hours) are pooled before estimating UR_C. Therefore, each sample for the system MSPI includes independent values of N for each of the component-type failure modes. Also, for systems with multiple trains containing similar components (e.g., the two motor-driven pump trains in an auxiliary feedwater system), the failures could occur randomly across the trains. Therefore, for the UA contribution to the MSPI, an average train FV/UA was used.

L.3   Simulation Results

For each sample in the overall simulation the individual plant system MSPI results were compared with the GREEN-WHITE threshold of 1.0×10^{-6}/year, WHITE-YELLOW threshold of 1.0×10^{-5}/year, and YELLOW-RED threshold of 1.0×10^{-4}/year. f the MSPI exceeded the thresholds, then it was tagged as WHITE, YELLOW, or RED as appropriate. Then the mean occurrence rate of each plant system MSPI being WHITE, YELLOW, or RED was determined over the 3,400 Latin hypercube samples. Summary results are presented in Table L.1. These results were obtained using the plant risk or SPAR model inputs (CDF_P, FV_P, UR_P, and UA_P), common-cause failure (CCF) impacts on FV/UR (multipliers presented in Appendix F), application of the frontstop, and application of the backstop.

Reviewing the first set of results in Table L.1, if the plant risk model inputs are used and CCF impacts are not included, then the fraction of WHITE MSPIs is 0.0337 (3.37 WHITES out of 100). This estimate corresponds well with the three MSPI WHITES observed in Appendix A (before the frontstop was applied). However, when the frontstop is applied, the fraction drops to 0.0203 (2.03 WHITES out of 100). Again, this estimate corresponds with (but is higher than) the one MSPI WHITE observed in Appendix A (after the frontstop was applied). When the backstop is also applied, the backstop provides an additional 0.0011 fraction of WHITES. Finally, the YELLOW fraction is 0.0007 and no RED occurrences were observed in the simulation.

If the MSPI is implemented as envisioned, the plant PRA results with CCF, frontstop, and backstop applied are most representative of expected future performance. In that case, the total WHITE fraction is expected to be 0.0285, the YELLOW fraction is 0.0016 (approximately one YELLOW for every 18 WHITES), and no REDS are expected. Again, these simulation results assume that the future performance of components would follow the baseline performance observed over the period 1999 – 2001.

Table L.1 Summary of MSPI Simulation Results

Result	Plant PRA		SPAR Resolution	
	no CCF	with CCF	no CCF	with CCF
WHITE Fraction	0.0337	0.0435	0.0390	0.0531
WHITE Fraction with Frontstop	0.0203	0.0275	0.0237	0.0312
% Reduction with Frontstop	40%	37%	39%	41%
Backstop WHITE Fraction	0.0015	0.0011	0.0014	0.0011
Total WHITE Fraction with Frontstop and Backstop	0.0218	0.0285	0.0251	0.0323
% of WHITES from Backstop	6.7%	3.7%	5.7%	3.4%
Additional YELLOW Fraction	0.0007	0.0016	0.0015	0.0023
Additional RED Fraction	0.0000	0.0000	0.0000	0.0000

APPENDIX M. THE SIGNIFICANCE OF COHORT EFFECTS

Appendix M
The Significance of Cohort Effects

M.1 Statement of the Issue

Because of a linear approximation, the MSPI tends to understate the change in *CDF* caused by redundant elements having simultaneous increases in unreliability or unavailability. The MSPI value increases linearly as basic event probabilities change; a more exact estimate of CDF grows more rapidly than linearly as basic event probabilities change. The present appendix discusses this "cohort effect" and discusses a potential approach for future consideration.

The point is illustrated in Table M.1. Suppose that there is a contribution to *CDF* from the product of two independent probabilities X and Y. Suppose further that both X and Y increase, causing an increase in *CDF*. Table M.1 compares the linear estimate of the induced change in *CDF* with the actual change in *CDF*.

Table M.1 Error in Estimated Changes in Contribution X*Y

% Change in X	% Change in Y	Actual % Change in X*Y	Linearized-Estimate % Change in X*Y	Error in Estimated Change
10	10	21	20	1/21
20	20	44	40	4/44
100	100	300	200	100/300

For small changes, the error is insignificant, but increases significantly as the magnitude of the change increases.

However, extrapolating this consideration to a realistic plant model is not straightforward. In practice, the CDF expression typically depends on X and Y not only through the product X*Y, as discussed above, but also through terms containing X (and not Y) as well as terms containing Y (and not X). Additionally, if X and Y are in the same common cause group, there is a CCF-related contribution to CDF which (in the SPAR model approach) is directly proportional to the probability of X Thus, the total change in CDF induced by concurrent changes in X and Y contains contributions that are NOT underestimated, summed with a contribution that IS underestimated, as summarized below in Table M.2.

Table M.2 Behavior of Contributions to Change in CDF when Two Elements Change Simultaneously

Terms Containing...	... Change In the Following Way When X and Y Change Together:
X (and not Y)	Linear (correctly estimated in linear approximation)
Y (and not X)	Linear (correctly estimated in linear approximation)
CCF (X, Y)	Linear (correctly estimated in linear approximation with the reservation that common cause parameters never change in the MSPI approach)
X*Y	Non-Linear (underestimated in linear approximation)

The above discussion has been carried out for contributions involving two variables. The effect is potentially stronger for contributions involving three or four variables appearing in the same cut sets.

M.2   Birnbaum Measure

The MSPI calculations in general, and the discussions provided in later subsections, are much easier to understand, given certain observations regarding "importance measures." This subsection provides those observations.

Suppose that we are interested in element A, and suppose that A appears in the *CDF* expression multiplying X and multiplying Y.[*] (*A* could be a basic event corresponding to unreliability or unavailability.) In addition to these *A*-related contributions, there are other contributions Z that do not contain A. Then we can write

$$CDF = A * X + A * Y + Z$$

The "Fussell-Vesely" "importance" FV of element A can be approximated[**] as

$$FV(A) = \frac{A * X + A * Y}{CDF}$$

This is the fractional contribution to *CDF* of terms containing A. The "Birnbaum" "importance" B of element A can be written

$$B(A) \equiv CDF(A = 1) - CDF(A = 0) = X + Y = FV(A) * \frac{CDF}{A}$$

Note that although the value of A appears in the denominator of the rightmost expression, the value of $B(A)$ is independent of the value of A, as shown in the preceding equality. For some purposes, it is useful to think of $B(A)$ as the partial derivative of *CDF* with respect to A. As illustrated below, $B(A)$ is the slope of a plot of *CDF* versus A. $B(A)$ is also the "coefficient" of A in the *CDF* expression.

Another noteworthy point to make about B is that elements that are logically in series have the same B (apart from considerations related to the caveats summarized at the end of this section), although they do not have the same *FV*. The reason that they have the same B can be seen most easily by substituting A1+A2 for A in the above discussion, corresponding to subcomponents A1 and A2 in series, and calculating the B of each.

---

[*] In this discussion, symbols such as "*A*" are being used to denote either a Boolean event or its probability, depending on context.
[**] The formulae in this section reflect the "rare-event" approximation. For present purposes, the effort needed to work around this approximation is not justified by the value added.

If performance changes in such a way that

$$A \to A + \Delta A,$$

then

$$CDF \to CDF + \Delta CDF = (A + \Delta A) * X + (A + \Delta A) * Y + Z$$

$$\Delta CDF = \Delta A * (X + Y) = \Delta A * B(A) = \Delta A * FV(A) * \frac{CDF}{A}.$$

The last line shows how to estimate changes in CDF without re-solving the entire model, once the "importance measures" B and FV have been obtained. This is how the MSPIs are evaluated: as if the B were constant. However, as illustrated below, the cohort effect causes the effective B of certain events to change along with the probabilities of those events.

The factor B(A), the coefficient of ΔA in ΔCDF, is essentially the context of A. If B(A) is overestimated, then the change in CDF will tend to be overestimated, and vice versa. The cohort effect is a concern in the MSPI implementation because the MSPI uses a constant (baseline) value of B(A), but B(A) itself can increase when the probabilities of multiple elements of a group change (as illustrated below). Thus, using a constant B(A) underestimates the change in CDF.

To see this in the algebra, consider simultaneous changes
$$A \to A + \Delta A,$$
$$X \to X + \Delta X$$
in the previous example. This leads to

$$CDF \to CDF + \Delta CDF = (A + \Delta A) * (X + \Delta X) + (A + \Delta A) * Y + Z$$

$$\Delta CDF = \Delta A * (X + Y + \Delta X) + A * \Delta X = \Delta A * (B_0(A) + \Delta X) + \Delta X * B_0(X).$$

In the last line, we have added a subscript 0 to B(A) to indicate that the value intended is the *original* Birnbaum, X+Y, not the new effective Birnbaum, X+Y+ΔX, and similarly for *B(X)*. In the last line, the change in Birnbaum is imputed to *ΔA*; clearly it could have been imputed to *ΔX* instead.

M.3   Examples

M.3.1 Overview

This section presents illustrative calculations performed with current SPAR models using SAPHIRE. These calculations illustrate points made above, and suggest an interim approach to addressing cohort effects in MSPI implementation.

M.3.2 Behavior of CDF with Only One Element Changing

As a calibration, consider first how CDF changes when only one element in the problem changes. Figure M.1 shows the change in CDF as a function of the change in the probability of an AFW turbine-driven pump's (TDP's) failure to start. This element was selected partly because it is not a member of a group, and therefore can change by itself, so there are no TDP-group cohort effects. Cohort effects across component types will be considered later.

In Figure M.1, the vertical axis is defined in terms of change in CDF relative to baseline. For convenience, the "hourly" CDF provided by SAPHIRE has been multiplied by 7000, to provide an estimate of CD events from full power per calendar year, for a plant having a capacity factor of 7000/8760. (This is the assumption used in MSPI work.) Also for convenience, the failure probability is presented in units of the baseline failure probability. Thus, "1" on the horizontal axis corresponds to the baseline value of TDP FTS probability (and, by definition, a ΔCDF of zero.) On the far right, the colored rectangles indicate the performance band implied by the corresponding change in CDF. The GREEN/WHITE threshold is at a ΔCDF of 1×10^{-6}. Thus, going to WHITE as a result of changing this element alone requires changing TDP UR by a factor slightly greater than 2.

As explained in Section M.2, the slope of this plot should be the Birnbaum. This element's baseline probability is 2.8×10^{-2}, and the baseline CDF in this model (in SAPHIRE hourly units) is 3.285×10^{-9}. Doubling the baseline probability yields a CDF of 3.437×10^{-9}. Thus, for a ΔUR of 2.8×10^{-2}, we get a ΔCDF of (3.437×10^{-9} - 3.285×10^{-9}), or 1.52×10^{-10}, implying a Birnbaum for this element of 5.429×10^{-9}. This agrees satisfactorily with the value printed by SAPHIRE for this element (5.423×10^{-9}).

Moreover, the constancy of the slope in this plot corresponds to the absence of a cohort effect for this case. The increase in CDF is simply proportional to the increase in TDP fail-to-start probability. The plot has been derived by explicitly requantifying the cut sets based on revised failure probabilities, but this linearity is exactly what the MSPI calculation also provides, and with the correct slope, as shown by the matching Birnbaum calculation.

M.3.3 Behavior of CDF with All Components of a Given Type Changing Simultaneously

Now consider changing the probability of failure to start of all of the diesel generators (DGs). As explained in the main body, this is in fact how the MSPI works: for a pool of like components (such as diesel generators), trials, successes, and failures are imputed to the pool, rather than to the individual components, and the pool-average value is then assigned to all members. When the members are mutually redundant (as the DGs are), they will appear in the same cut sets, at least some of the time, giving rise to a cohort effect.

Figure M.2 shows how CDF changes with increasing DG failure probability. The conventions are as above: the horizontal axis is presented in units of the baseline event probability, and the vertical axis is defined in terms of change in CDF relative to baseline, for 7000 hours per year of full-power operation.

The top curve is obtained from requantifying the cut sets as failure probability changes. Within this SPAR model, changing this failure probability also changes a common-cause contribution. This effect is included in the top curve. The nonlinearity of the top curve signals a cohort effect within the pool of diesel generators.

The bottom curve is the linearized estimate derived from multiplying ΔUR (for the fail-to-start failure mode) by the individual DG Birnbaums, and adding them up. The middle curve additionally includes the contribution from common cause, which is not reflected in the individual DG Birnbaums.

The gap between the top curve and the middle curve represents the importance of the cohort effect. We see that for small scale factors, the error is not too great, but for larger scale factors it is significant. In this example, the neighborhood of the GREEN/WHITE threshold is not affected very strongly, but the transition from WHITE to YELLOW is significantly affected.

Figure M.3 looks at the cohort effect in another way. The Birnbaum measure is plotted versus scale factor for two elements: the TDP failure to start considered in the previous subsection, and one of the diesel generators' failure to start. In this figure, the TDP Birnbaum is plotted as a function of TDP changes (with DG fixed), and the DG Birnbaum is plotted as a function of DG changes (with TDP fixed); effects across component types are addressed later. As implied by Figure M.1, the TDP FTS Birnbaum is constant as TDP FTS probability changes. The Birnbaum of an element measures what is in parallel with the element, and is necessarily constant for the TDP element if the only thing changing is the TDP itself. The diesels, on the other hand, are in parallel with each other, and change in probability as a group; therefore, their Birnbaum measures also change, as illustrated on this figure.

M.3.4 Behavior of CDF with Multiple Component Types Changing

In previous subsections, only one component type was changing at a time. Here, simultaneous changes in TDP performance and in DG performance are considered. Many variations can be postulated: the same scale changes in both, a large change in one and a small change in the other, and so on. Figure M.4 shows changes in CDF as a function of a scale factor applied simultaneously to TDP failure to start and to DG failure to start, and changes in CDF for each of these changes applied separately.

It has already been shown that there is a cohort effect within the diesels. If there were no cohort effect between the TDP and the diesels, then the concurrent change would be the sum of the DG and TDP changes. On the far right of the graph (Scale Factor = 10), the concurrent change is seen to be greater than the sum of the independent changes. Therefore, there is a cohort effect between the TDP and the DGs. This is expected, and the example was chosen based on this expectation; the TDP has an important role to play when the DGs fail.

Figure M.5 shows an expanded view of Figure M.4, focusing in the region of the GREEN/WHITE threshold. In this region, the cohort effect between TDP and DGs is seen to be not too significant: the concurrent change is approximately equal to the sum of the independent changes.

M.3.5 Cohort Effect between Fail-to-Start and Test & Maintenance

The plant analyzed above contains three diesels, and in many scenarios needs two for success. Thus, the leading cut sets involving diesels involve either a common cause event, or two DG-related basic events. Frequently, the two-DG events are a test or maintenance unavailability in conjunction with a failure to run or a failure to start. Such cut sets will not manifest a cohort effect if only the fail-to-start probability changes. Figure M.6 shows the effect of changing only T&M, changing only FTS, and changing the two concurrently. As suggested by the observation that many leading cut sets contain a T&M and an FTS, there is a comparatively large cohort effect from this conjunction of T&M and FTS. It should be added that the changes postulated in this example are very large.

Note that in Figure M.6, the scale factor is only going out to 7, and that the vertical scale is relatively compressed owing to the rather large number obtained for the concurrent-change case. These two differences make the FTS-change curve look less nonlinear than it did on earlier figures. However, it is interesting to compare the FTS-change curve with the TM-change curve, which really IS linear. There is no intra-TM cohort effect, because multiple-DG-TM events do not appear in the cut sets, presumably for reasons related to technical specifications.

M.3.6 Cohort Effect in "Insensitive" System

In the cases examined above, by the time the cohort effect became very significant, the performance was already solidly "WHITE." However, those systems were relatively important; one can ask how the behavior differs for systems that require large-scale changes to go WHITE at all. Figure M.7 shows results for a relatively insensitive system (RHR at a boiling-water reactor, or BWR). By "insensitive," we mean in this case that it takes a six-fold increase in failure to start (acting by itself) to cause a CDF change of more than 1×10^{-6}. (For comparison, the diesels get there with a factor of 3 at this plant, and a factor of 2 at the other plant analyzed.)

Figure M.7 shows results for increasing FTS (with TM fixed), increasing TM (for FTS fixed), increasing FTS and TM simultaneously, and (for convenience) the sum of FTS (with TM fixed) and TM (with FTS fixed). The gap between the latter two curves represents the cohort effect between FTS and TM. As for a more sensitive system illustrated earlier, there is a significant cohort effect between FTS and TM. The slight curvature in the FTS-only plot reflects the slight cohort effect of FTS ANDed with FTS events, and the lack of curvature in TM-only implies (as before) that concurrent TM on RHR pumps is not a contributor.

Neither change by itself (FTS or TM) exhibits a significant cohort effect. In the expression analyzed, there are 585 cut sets containing at least one FTS event; but only 28 of these contain two FTS events. There are many terms containing an FTS event from one train, and TM or non-pump components from the other.

In this case, the cohort effect between TM and FTS is manifest by the time the change in CDF reaches 1×10^{-6}. Therefore, the cohort effect could, in principle, cause a MSPI "GREEN" to be declared where a fully model-based calculation would declare a "WHITE." Even in this case, the window of opportunity for that outcome is relatively narrow (it appears to be a very special case), at least in scale-factor space.

M.4 How Often Will The Cohort Effect Be Significant?

The magnitude of the cohort effect is example-specific. While it can be large for relatively large changes in basic event probability, it is not necessarily very large in the neighborhood of the GREEN/WHITE threshold. In several special cases examined, for very large changes in basic event probability (factor of 10), the underestimate of the linear approximation is on the order of 50%. For "sensitive" systems, where small changes cause the indication to go WHITE, these large errors occurred when performance was already "WHITE," so a GREEN indication would probably not have resulted from the error. For "insensitive" systems, where relatively large changes are needed to get to the GREEN/WHITE interface, the situation is potentially more complicated. In the "insensitive" case examined above, the effect was indeed somewhat worse at the GREEN/WHITE interface than it had been for the sensitive case. Although the change in CDF is small at the GREEN/WHITE interface, it is caused by relatively large changes in element probabilities, and this is the condition under which the cohort effect is potentially significant.

How often will changes of this magnitude occur in practice? Figure M.8 shows the measured probability of exceeding a given scale factor, based on recent data. This figure pools data for all component types, so the indicated probability of exceedance is not necessarily typical of a given component type. However, it suggests the following:

- Large scale factors are not typical.
- However, scale factors greater than 3 can be observed more than one percent of the time.
- Significantly higher scale factors are not rare events.

Much of the time, the scale factors are low, so the cohort effect is insignificant. However, some of the time, the scale factors are not low enough to justify neglect of the cohort effect.

M.5 Addressing the Cohort Effect Computationally

M.5.1 Improved Approximation

Recognizing that the MSPI approximation is based on the linear terms in the expansion of ΔCDF, it is natural to ask whether the issue can be addressed by going to the next order terms.

When a single parameter (such as the probability of DG failure to start) is changing (affecting multiple redundant elements), the second-order Taylor approximation of the effect on CDF is

$$CDF(A_0 + \Delta A) \approx CDF(A_0) + \left.\frac{\partial CDF}{\partial A}\right|_0 \Delta A + \frac{1}{2}\left.\frac{\partial^2 CDF}{\partial A^2}\right|_0 (\Delta A)^2 \ .$$

It follows that

$$\Delta CDF \approx \left.\frac{\partial CDF}{\partial A}\right|_0 \Delta A + \frac{1}{2}\left.\frac{\partial^2 CDF}{\partial A^2}\right|_0 (\Delta A)^2 \ .$$

$$= B\Delta A + \frac{1}{2}B' \cdot (\Delta A)^2$$

The first derivative of CDF is the Birnbaum importance, *B*. The second derivative is the slope of the Birnbaum importance, *B'*.

The MSPI approximation to the change in CDF corresponds to the first term. For example, Figure M.3 shows two cases: one in which *B'* is zero, and the MSPI approximation is correct, and one in which the Birnbaum importance rises essentially linearly, so that *B'* is essentially constant but clearly different from zero. For this case, the second-order formula gives the result shown in Figure M.9. The "CDF" series plots the same CDF calculations as those shown in Figure M.2; the "Second-order" and "Linear" plots are, respectively, based on the second-order approximation shown above, and on the first-order term of it. One sees that the second-order approximation is very good in this range. In this case, the first and second derivatives were derived from three CDF quantifications performed at scale factors of 0, 1, and 2; this process (although numerically crude) automatically captures the change in CDF occasioned by the common cause model and all group importance effects. Approximations to the first derivative could have been obtained from importance measures for a single model quantification, but another run would have been necessary to get the second derivative.

In examples where the cut sets contain no terms of order higher than A^2, the second-order Taylor approximation is exact. In examples with higher-order terms, the second-order approximation is still superior to the first-order (i.e., constant Birnbaum) approximation, and may be fully adequate in the range of ΔA of interest.

Figure M.10 shows a calculation similar to the above for DGs, but for a different plant (one with four diesels). Here, the second-order correction is a big improvement, but a gap remains. This reflects the circumstance that some cut sets contain more than two FTS events.

M.5.2 Implementation Issues

If the cohort effect only mattered for single component types, one could address it simply by capturing the single-component second derivative effects in the data entry file. Unfortunately, cross-terms also need to be considered. Suppose that two distinct basic-event probabilities are considered, *A* and *C*. (We reserve the letter *B* for Birnbaum.) Then the Taylor expansion is

$$CDF(A_0 + \Delta A, C_0 + \Delta C) \approx CDF(A_0, C_0) +$$

$$\left.\frac{\partial CDF}{\partial A}\right|_{0,0} \Delta A + \left.\frac{\partial CDF}{\partial C}\right|_{0,0} \Delta C +$$

$$\frac{1}{2}\left.\frac{\partial^2 CDF}{\partial A^2}\right|_{0,0} (\Delta A)^2 + \left.\frac{\partial^2 CDF}{\partial A \partial C}\right|_{0,0} \Delta A \Delta C + \frac{1}{2}\left.\frac{\partial^2 CDF}{\partial C^2}\right|_{0,0} (\Delta C)^2$$

Denote the Birnbaum importance of *A* by B_A, and use similar notation for the other terms. Then, the above formula can be written as follows:

$$\Delta CDF \approx B_A \Delta A + B_C \Delta C + \frac{1}{2}\left(B_A\right)'(\Delta A)^2 + \left.\frac{\partial^2 CDF}{\partial A \partial C}\right|_{0,0} \Delta A \Delta C + \frac{1}{2}\left(B_C\right)'(\Delta C)^2 .$$

This requires a new piece of information: the cross-term second derivative. Doing the MSPI for a given system would, in principle, require evaluating the cross-term for all pairs of component types within each system. In practice, some pairs might be negligible.

Once the second derivatives are obtained and put into data entry files, implementation would require no more inputs than it does today; the MSPI calculation would take changes in UR and UA as inputs, and furnish index values as outputs. The illustrations provided above suggest that doing this would provide a significant improvement for large scale factor changes in the inputs. However, it would require some effort at the outset to calculate all the numbers, and it should be pointed out that even this approach would neglect cross-system cohort effects.

M.6 Summary

When multiple basic events' probabilities increase, the linearized MSPI estimate of the resulting change in CDF is an underestimate, if individual cut sets contain more than one of the changing basic events.

This occurs because in the linearized MSPI calculation, the impact of each basic event's change is quantified as if all the other event probabilities remained constant. In general, this assumption is not satisfied. For example, redundant similar components appear together in cut sets, giving rise to cohort effects even when only that component type changes. Additional cohort effects may occur if different component types change simultaneously, or if different failure modes or unavailability contributors change for a given component type.

If one focuses only on cut sets containing many elements that are concurrently changing, and additionally focuses on increases that are large relative to baseline values, one might suppose that the linearized approximation cannot be very good. However, for changes that are not relatively large, the approximation is not bad in many cases. This result appears to be related to the circumstance that because of the structure of the problem, some significant contributors to ΔCDF are, in fact, linear. That is, in many cases, even when multiple elements are changing, there is a significant (linear) contribution from cut sets containing individual changing elements, but NOT containing conjunctions of the concurrently changing elements. This is exemplified in the DG cases shown above, where even though redundant elements' failure probabilities are concurrently changing, the cohort effect is not necessarily great. However, the DGs also offer a counterexample: when both fail-to-start and maintenance increase together, a significant cohort effect is seen, because there is a very significant contribution from conjunctions of this type.

In a sense, the cohort effect between FTS and TM only matters if significantly declining performance in the two areas is a real possibility. In principle, assessing the significance of the cohort effect between FTS and TM requires us to formulate the prior probability of the occurrence of increases in both of these event types (failure events and maintenance outages). An increase in the number of failures might be expected to force an increase in corrective maintenance; on the other hand, an increase in preventive maintenance might reduce the failure count. Formulating prior positions on matters such as this (the prior likelihood of any given postulated cohort change) is beyond the scope of the present appendix.

Formally, one option would be simply to go to a model-based (rather than linear-approximation-based) quantification of CDF changes with changing performance. This has significant programmatic drawbacks that are beyond the scope of the present discussion. It is desirable to retain the simplicity of the MSPI approach as much as possible.

A possible approach that would retain much of the simplicity of the current MSPI approach would be to add second-order correction terms to the MSPI formula. If cut sets contain at worst quadratic combinations of varying elements, this approach could provide numerically exact results to the change in CDF (if all of the cross-terms are captured), and even if higher-order combinations appear in cut sets, the second-order correction is a big improvement. Implementing it would require some startup effort to generate the necessary second derivatives and implement the calculation in the data entry files.

Short of addressing the cohort effect computationally as discussed above, it is difficult to articulate a rule of thumb that will be simple to understand, efficient in application, and guaranteed not to make any classification errors. Moreover, arguing the efficacy of such an approach would require evaluating the prior likelihood of particular cohort changes, a difficult discussion to justify at present.

This suggests the approach of closely observing the MSPI calculation during the initial implementation phase to see how well it does in practice. The examples discussed above suggest that the cohort effect seldom makes much difference at the GREEN/WHITE threshold, or when scale factors are less than two. At the WHITE/YELLOW threshold, however, it makes a significant difference. This suggests that when *multiple* basic event probabilities are changing by factors of more than 2, or the estimated change in CDF is well in excess of 1×10^{-6}, it may be worth propagating the changes through the full model simultaneously, to see whether the declaration ought to be YELLOW rather than WHITE, or, in rare cases, WHITE rather than GREEN.

It should be kept in mind that the MSPI is primarily to be used as an indicator of deviation of system performance from the norm. As such, it is recognized that the MSPI is simply a measure of the change of system unavailability and component unreliability from some historical baselines, where the various components are weighted by their relative risk importances. For certain applications such as online risk monitoring or technical specification changes, the linear approximation used in the MSPI would be inappropriate. This is especially the case when the removal from service of certain high risk-important components could cause increases of factors of 2 to 10 in CDF. For changes in CDF of the magnitude of 1×10^{-7} to perhaps mid-1×10^{-5}, the linear approximation is generally satisfactory for what is intended — namely, to cause increased attention (in the form of increased inspection resources) to be focused on the appropriate system and components.

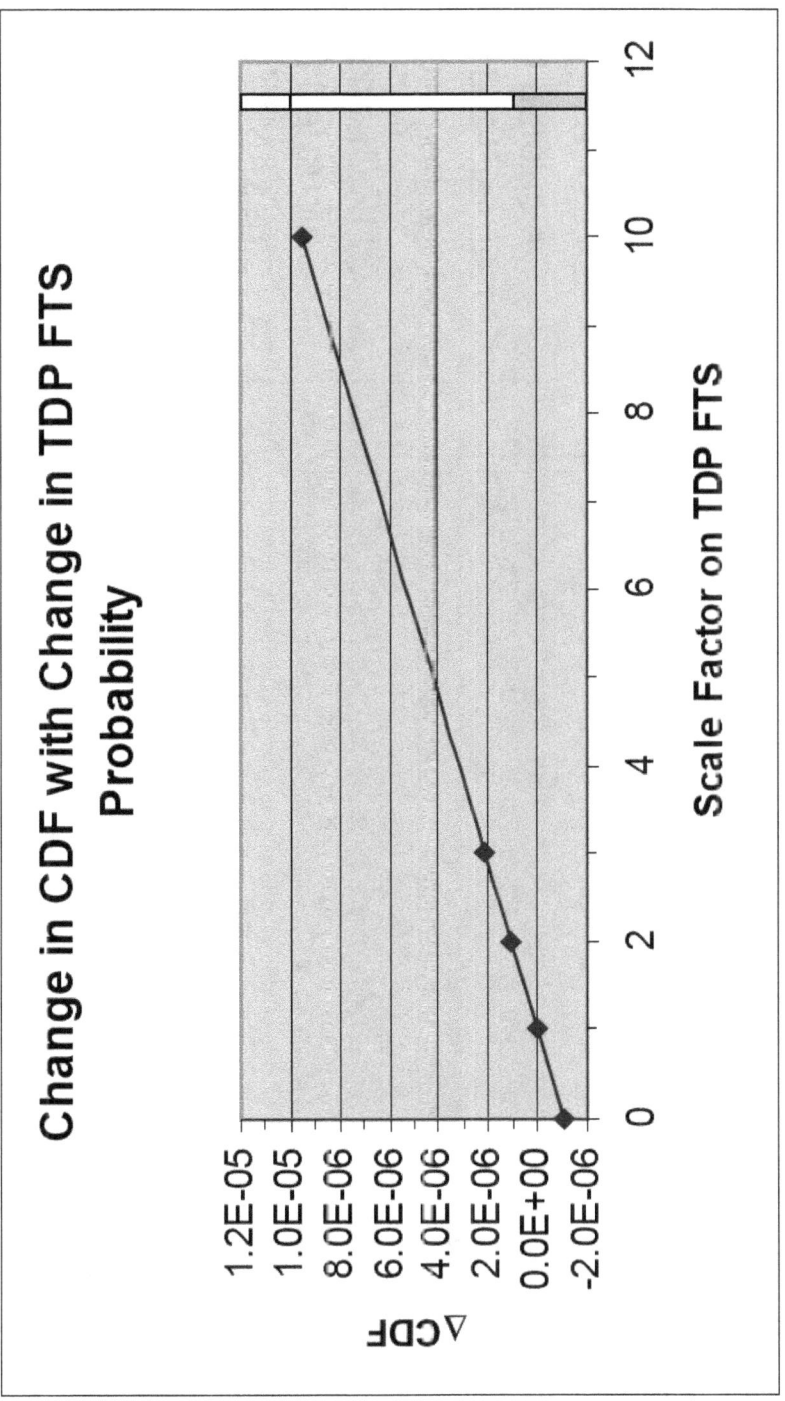

Figure M.1 Change in CDF with Change in TDP FTS Probability

M-11

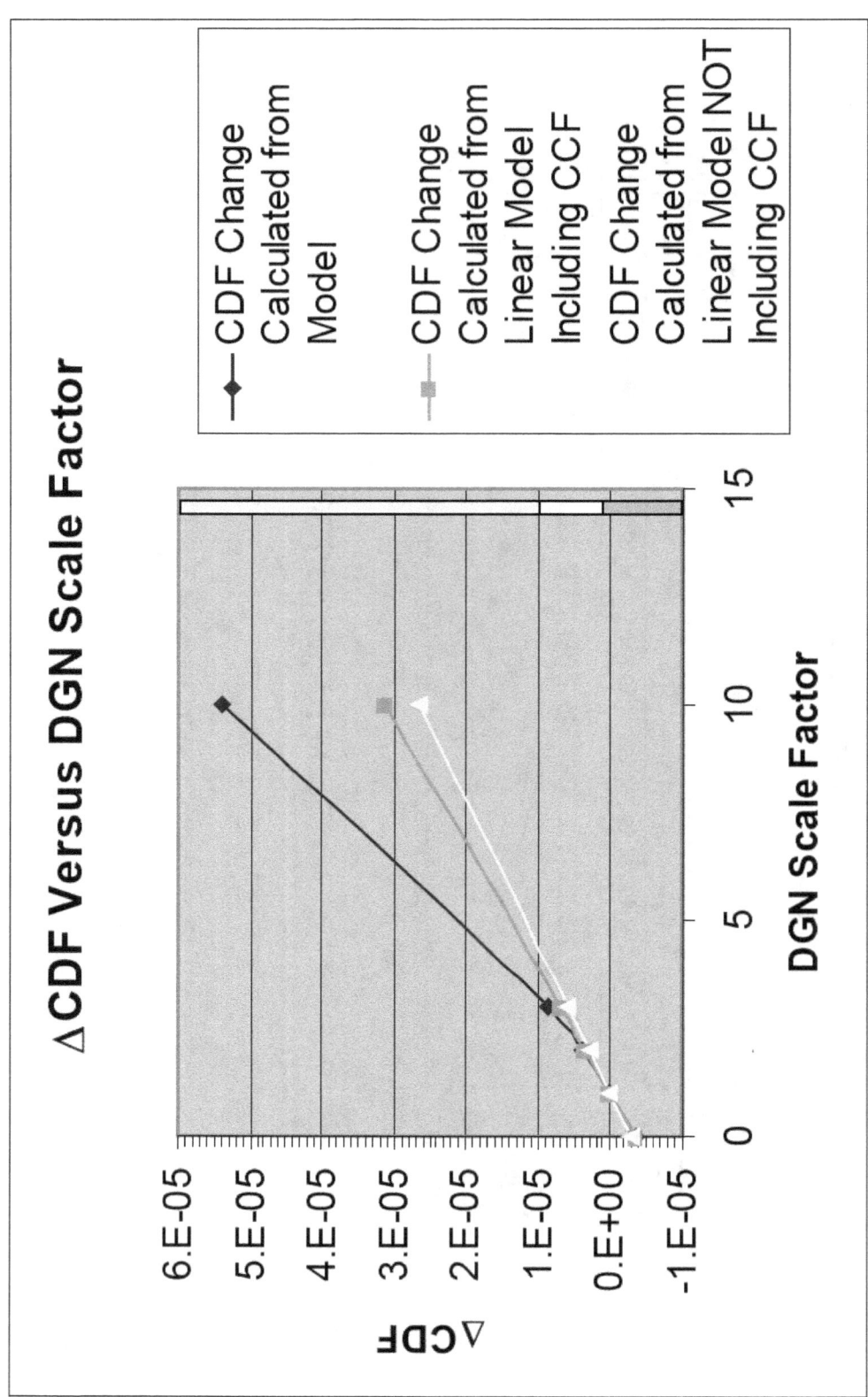

Figure M.2 Change in CDF with Change in DG Fail-to-Start Probability

M-12

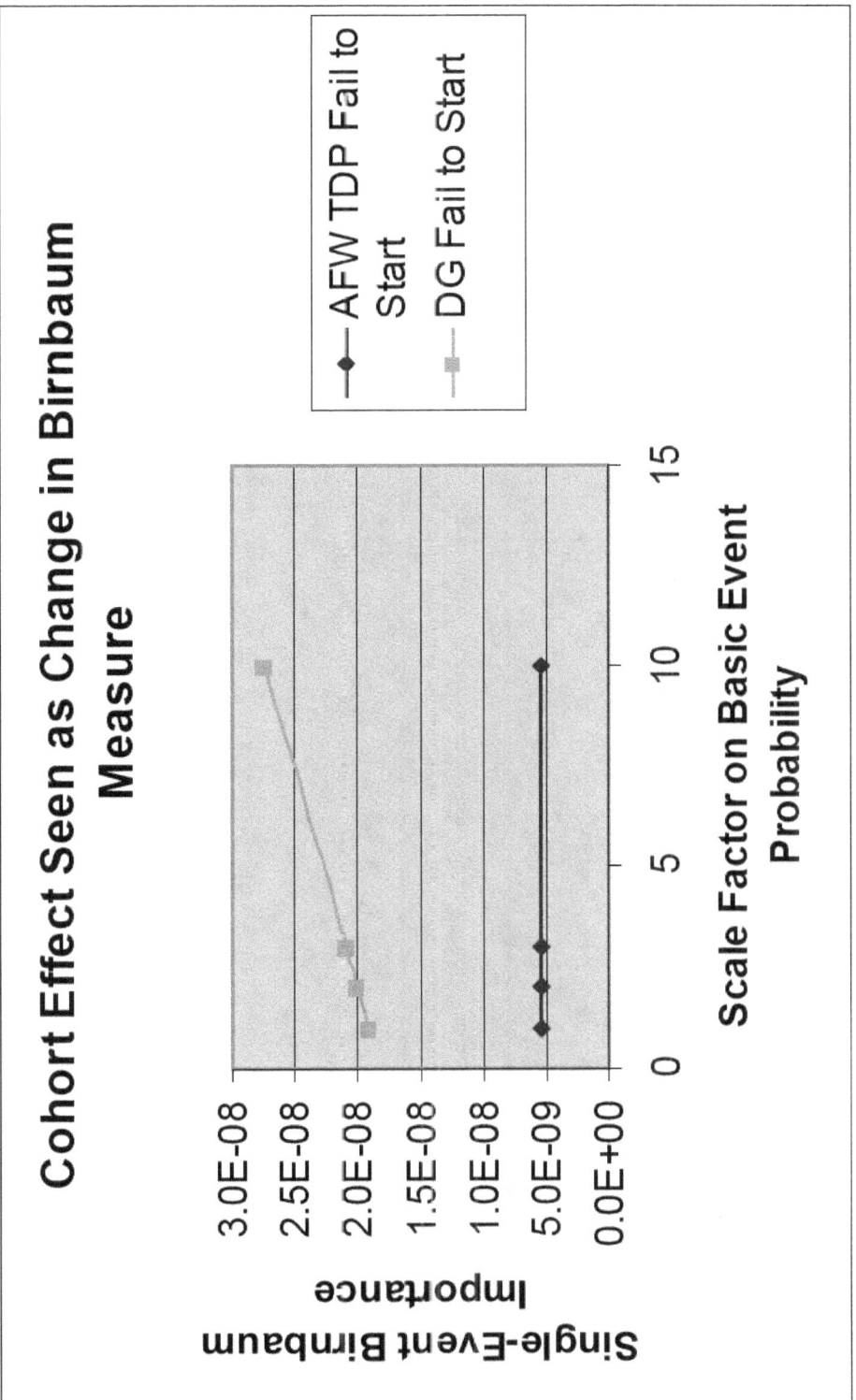

Figure M.3 Change in Event Birnbaum with Change in Event Probability

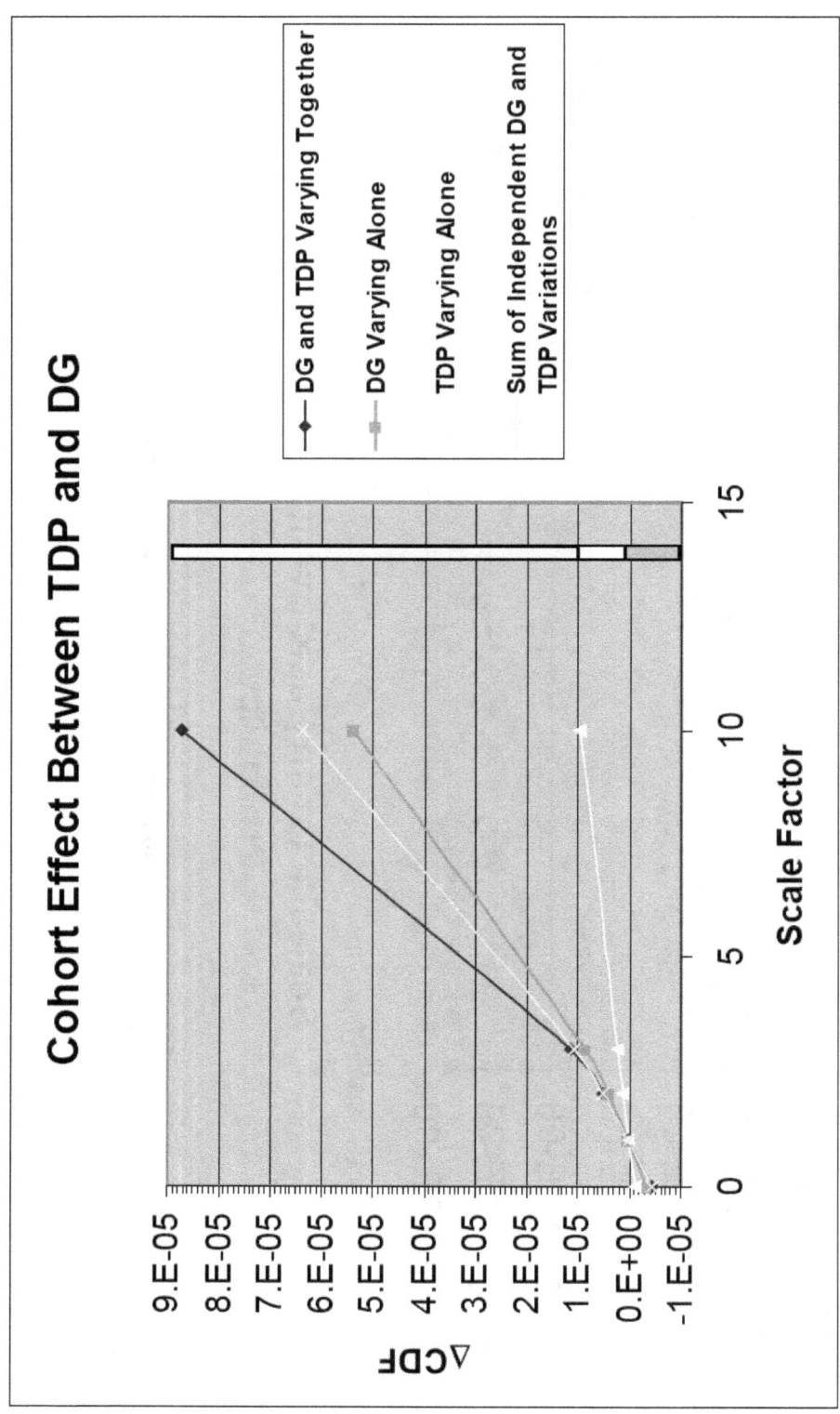

Figure M.4 Change in CDF with Concurrent Changes in TDP and DG FTS Probabilities, Compared with TDP Alone and DG Alone

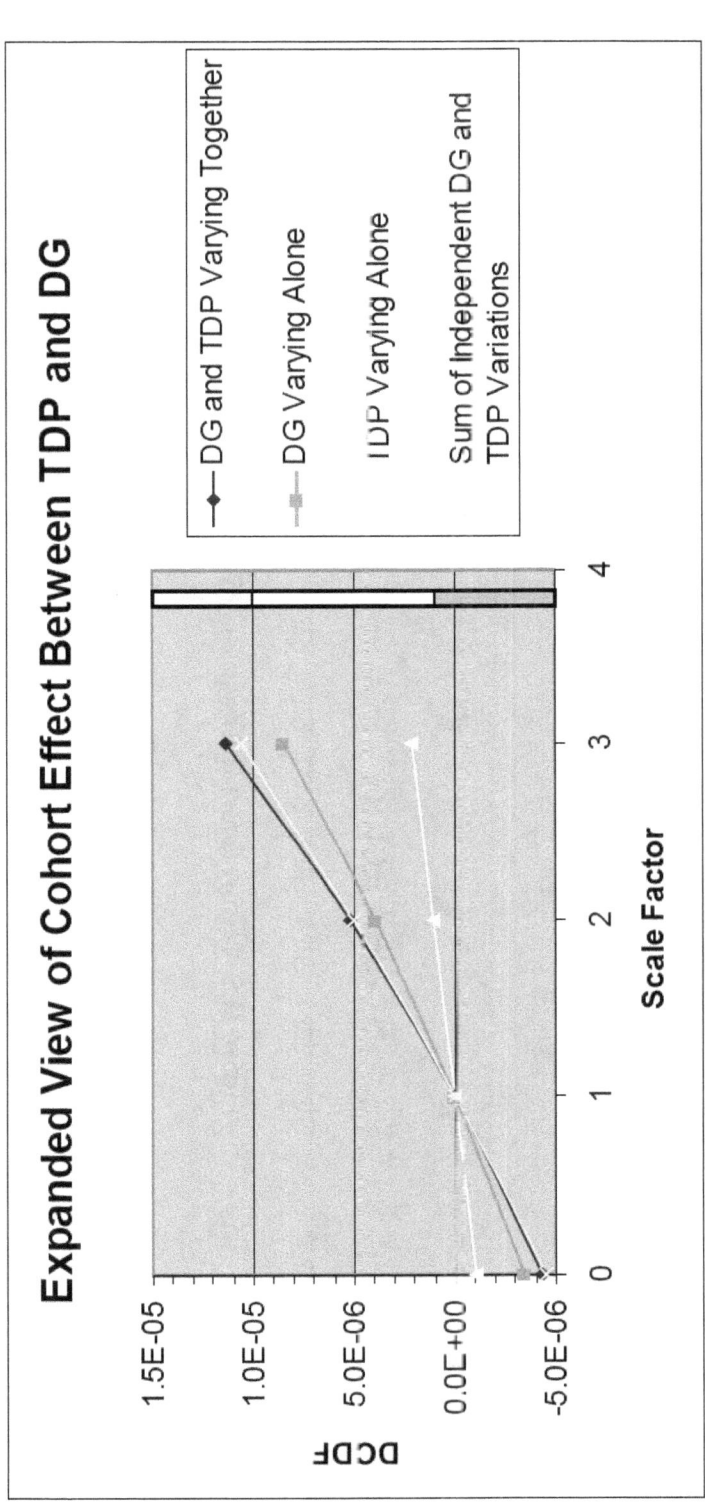

Figure M.5 Expanded View of Figure M.4, Showing That in This Range, "Concurrent" Result Is Approximately the Sum of the Independent Results

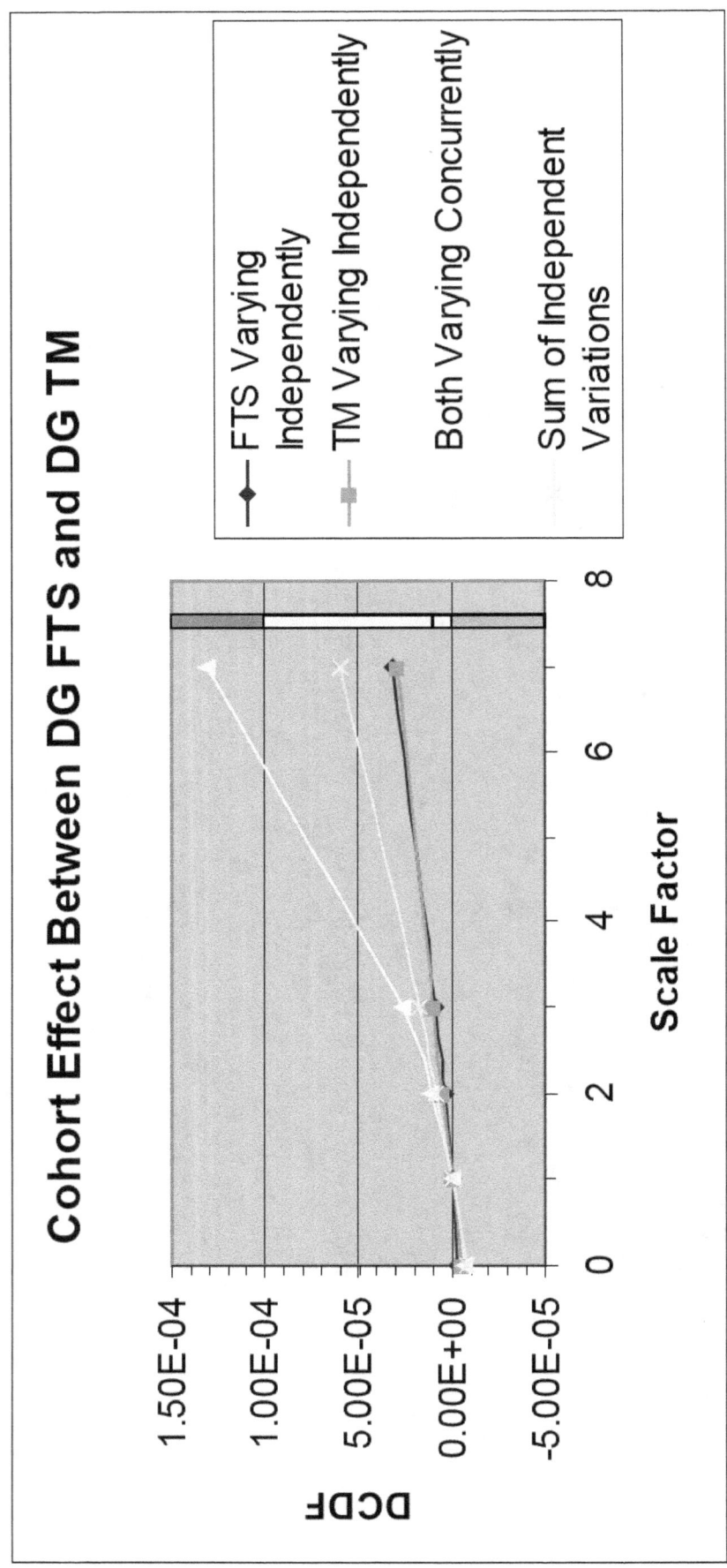

Figure M.6 Cohort Effect Between DG FTS and DG TM

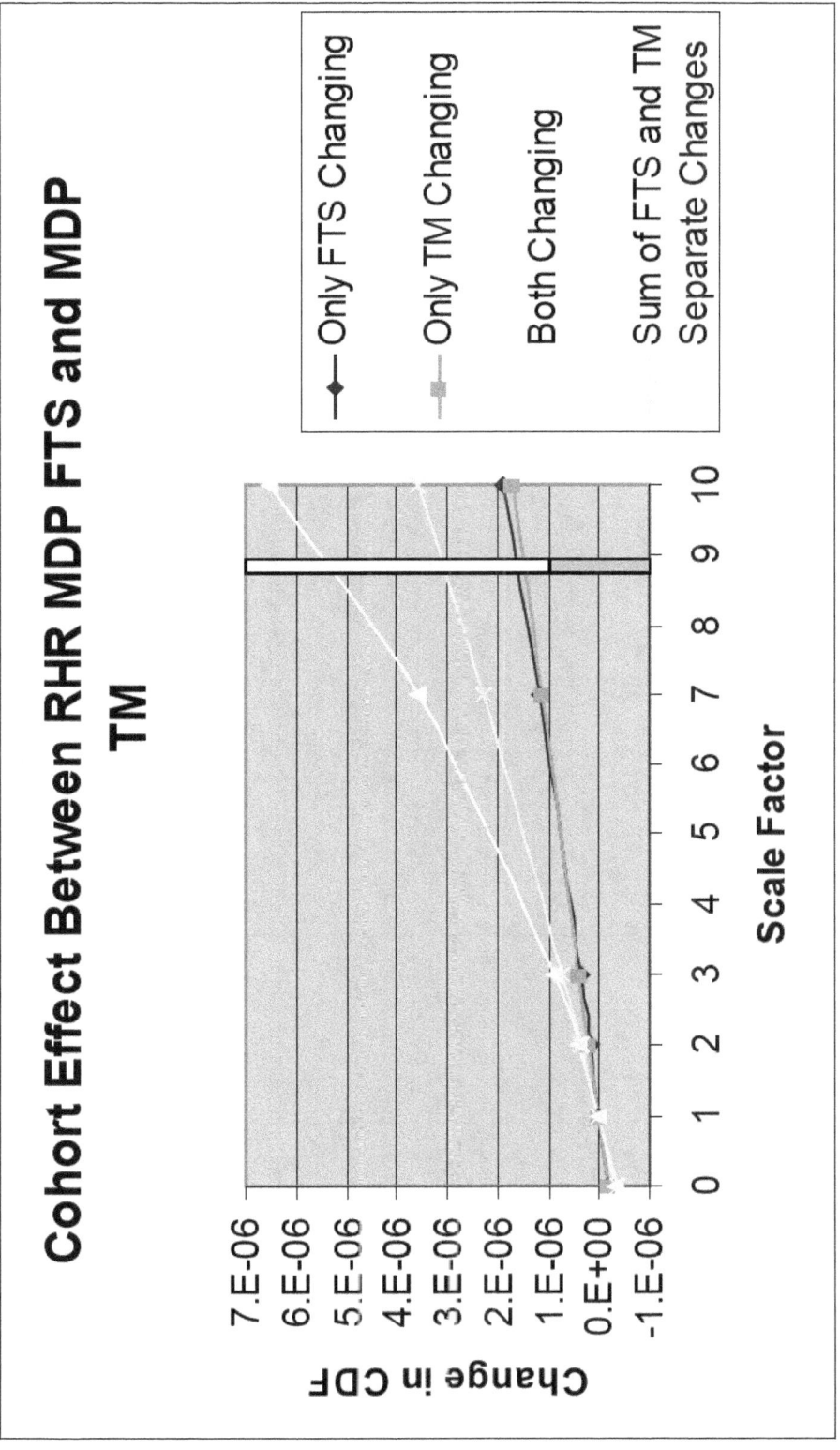

Figure M.7 Change in CDF for a Relatively "Insensitive" System

MSPI Pilot Plant Component Type UR Scale Factors
4th Quarter 2002

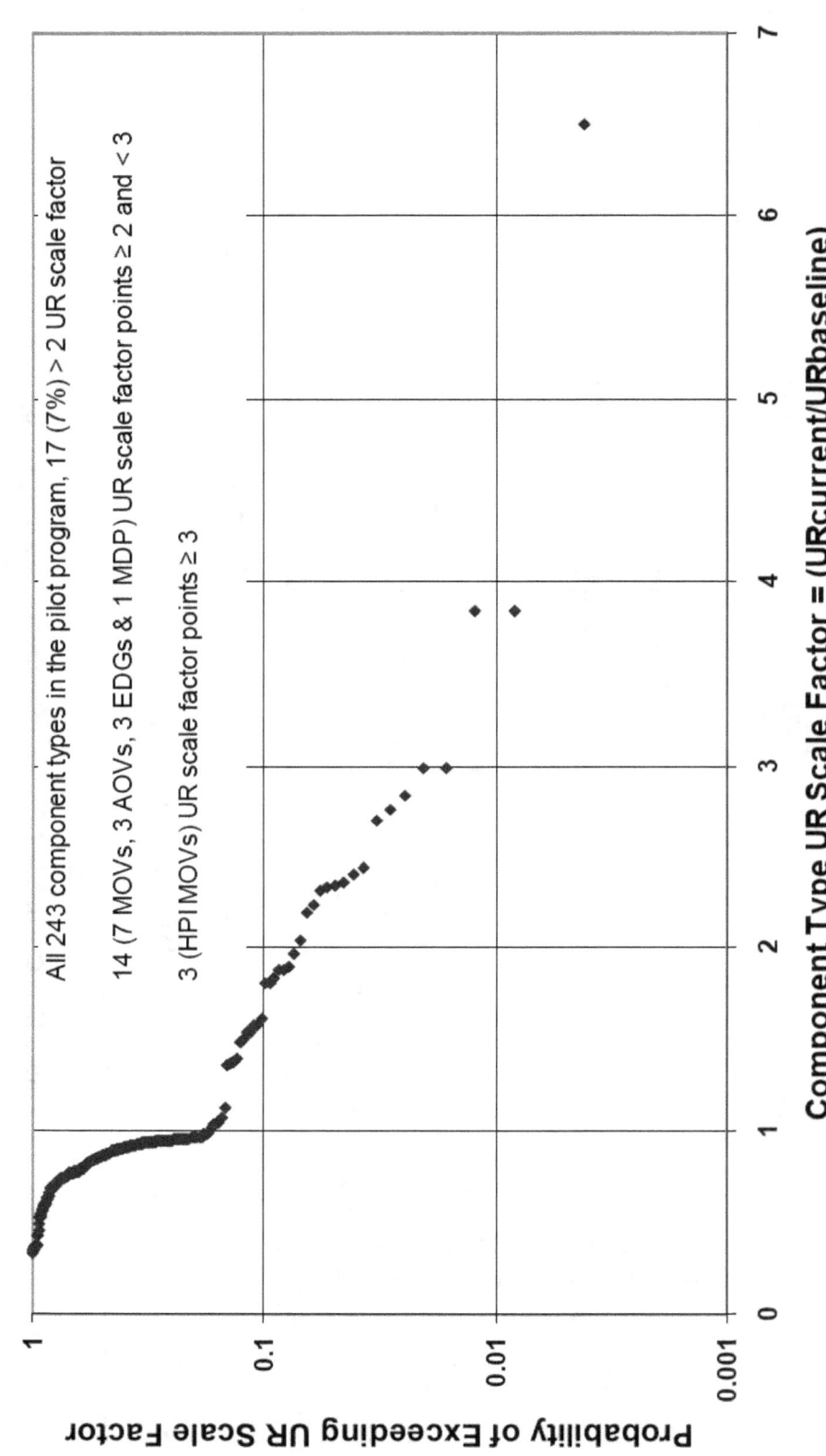

All 243 component types in the pilot program, 17 (7%) > 2 UR scale factor

14 (7 MOVs, 3 AOVs, 3 EDGs & 1 MDP) UR scale factor points ≥ 2 and < 3

3 (HPI MOVs) UR scale factor points ≥ 3

Component Type UR Scale Factor = (URcurrent/URbaseline)

Probability of Exceeding UR Scale Factor

Figure M.8 Observed Fraction of Components Exceeding UR Scale Factor In A Recent Observation Period

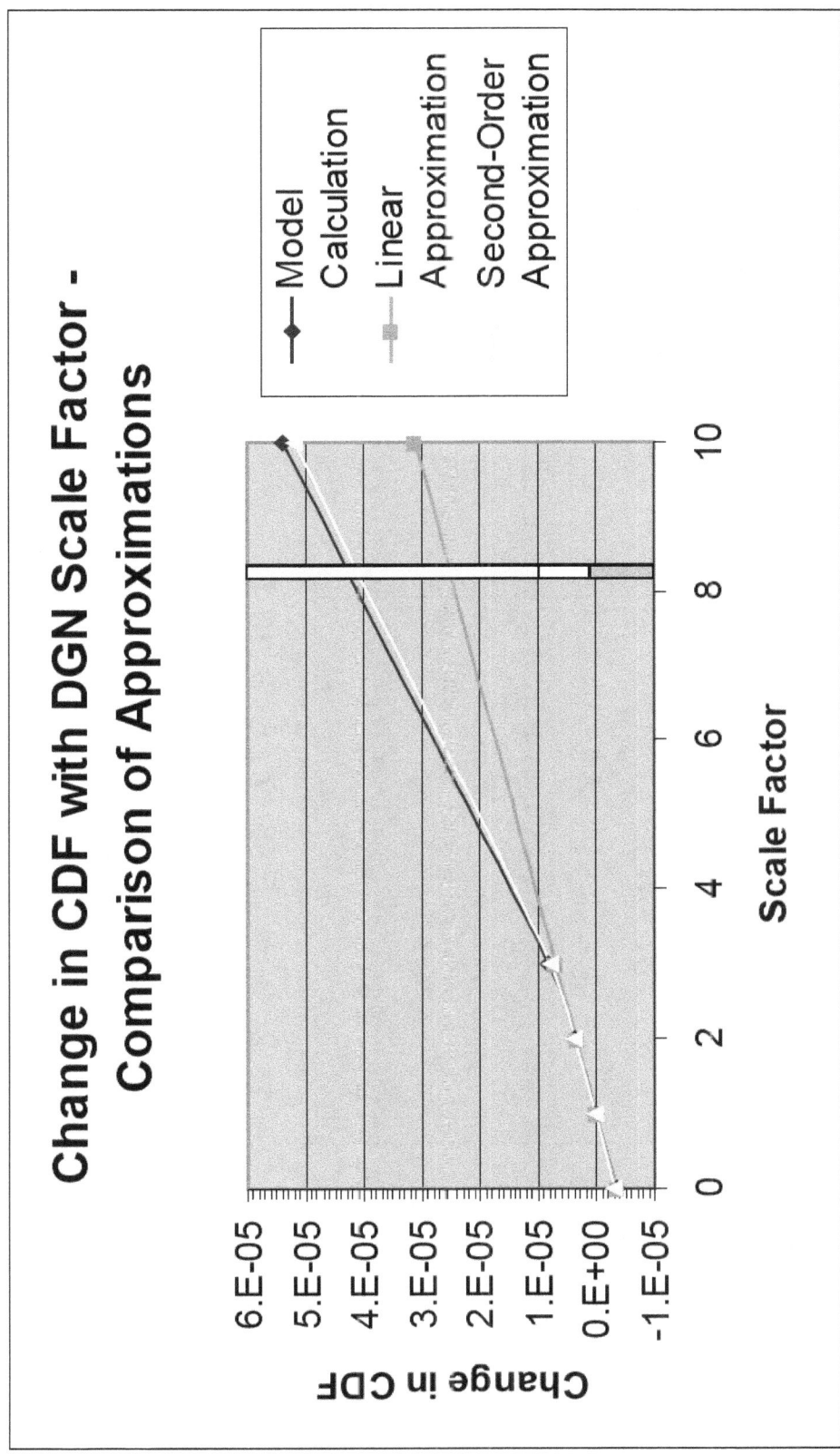

Figure M.9 Comparison of Model Calculations with First- and Second-Order Estimates of Change in CDF for a 3-DG Plant

M-19

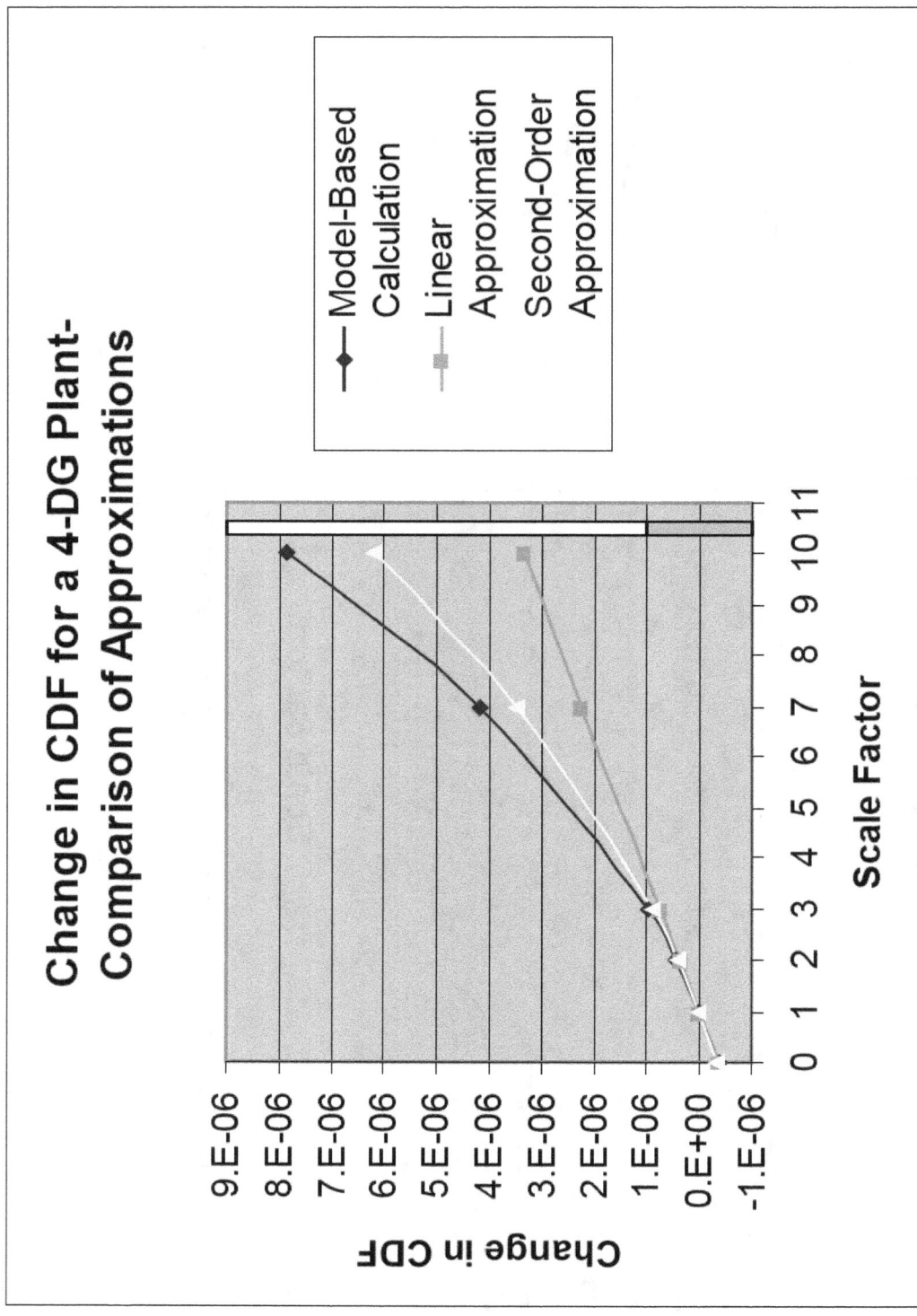

Figure M.10 Comparison of Model Calculations with First- and Second-Order Estimates of Change in CDF for a Four-DG Plant

M-20

APPENDIX N. MSPI PUBLIC REVIEW COMMENTS

Appendix N
MSPI Public Review Comments

Table N.1 presents the NRC's responses to comments received from external stakeholders during the review of the *Report on the Independent Verification of the MSPI Results for the Pilot Plants*. In a notice in the *Federal Register*, 69 FR 20953, dated April 19, 2004, the NRC requested public comments on the MSPI verification report. The due date for comment submittals was June 15, 2004. Thank you to the individuals who submitted comments:

F.G. Burford, *Entergy*
Mark Burzynski, *TVA*
Fred Madden, *TXU Power*
L. William Pearce, *FENOC*
Anthony Pietrangelo, *NEI*
Bill Vesely, *NASA*
D.R. Woodlan, STARS
Mario V. Bonaca, ACRS

Table N.1 Public Comments on the Review of the MSPI Verification Report and Responses

	Comment	Source	Response	Reference
1.	*General Comment.* We believe that the subject report provides an accurate formulation of the analyses, decisions and consensus developed over the past several years and should be used in revising the proposed MSPI guidance.	NEI	None required.	N/A
2.	*General Comment.* We believe the fundamental mathematical formulation of the MSPI is an appropriate, simplified indication of the net change in core damage frequency for chosen systems at an individual plant based on system unavailability and component unreliability compared to the industry baseline. With the changes proposed in the report's recommendations, we believe that the MSPI will provide a robust indication of performance far superior to the current SSU indicator.	NEI	None required.	N/A
3.	*General Comment.* Industry supports the overall technical findings of the report. The validity and robustness of MSPI outcomes are accurately and fully described in Appendix I, MSPI/SSU/SDP Benchmark." We concur with the concluding statement," .... the MSPI appears to consistently provide the best overall measure of integrated system performance, while minimizing both false positive and false negative likelihoods."	NEI	None required.	N/A
4.	*Comments on Recommendation #1.* Industry accepts this recommendation with the previously stated concern that the values should be revalidated when the entire industry has provided the 3 years of historical data.	NEI	None required.	Executive Summary
5.	*Comments on Recommendation #2.* Industry accepts this recommendation. Without the "frontstop," as the report details, there would be a significant number of false positives, which do not reflect licensee performance. Performance of a Significance Determination Process analysis by the NRC staff will provide a different assessment of the single failure.	NEI	None required.	Executive Summary
6.	*Comments on Recommendation #3.* Industry accepts this recommendation. A high number of component failures, even if less risk significant in total, can signal a decline in performance outside the industry norm.	NEI	None required.	Executive Summary

	Comment	Source	Response	Reference
7.	*Comments on Recommendation #4.* Industry accepts this recommendation; however, we believe further work is warranted to ensure success in implementation. The generic factors derived and published in the report should be applied to all plants and systems so that there will be no confusion as to which factors apply. Thus, a table similar to Table F.4, Recommended Generic CCF Multipliers by Pilot Plant, should be expanded to cover all plants. Prior to implementation, licensees can individually review and comment if necessary on the appropriateness of the factors applied to their components.	NEI	We agree that Table F.4 should be expanded to include CCF multipliers for all plants prior to implementation.	Executive Summary
8.	*Comments on Recommendation #5.* Industry accepts this recommendation.	NEI	None required.	Executive Summary
9.	*Comments on Recommendation #6.* Industry accepts this recommendation.	NEI	None required.	Executive Summary
10.	*General Comment.* We endorse the comments provided by NEI.	Entergy, FENOC, TVA, STARS, and TXU Power	None required.	N/A
11.	*General Comment.* The general MSPI methodology is valid. The above report does a commendable job in carrying out comprehensive studies to verify the applicability of MSPI. Since significant effort has been expended on verifying the MSPI for the cases studied in the report, I will focus on areas that are not covered and that can impact or enhance the MSPI. These areas involve 1) presenting the basis for the MSPI to show its domain of applicability and its expansion potentials, 2) defining explicit criteria that determine when the MSPI is applicable and when it is not, 3) correcting for the inertia in the MSPI that can cause erroneous results, and 4) providing an alternative, simpler notation for the MSPI to enhance its understandability, and 5) explicitly quantifying the uncertainty in the MSPI. Each of these areas is covered below.	Bill Vesely	None required.	N/A

	Comment	Source	Response	Reference
12.	Multiple unavailability and unreliability changes of *cohort* equipment (i.e., equipment in the same minimal cutset) can be significant. The report needs to address this. It should formulate the MSPI in terms of the Taylor expansion of CDF, and link the current formulation to the linear approximation. It should establish the domain of validity of the MSPI (the magnitude of performance changes for which the MSPI is valid), and when the linear approximation is not valid, the MSPI should append the appropriate higher-order corrections.	Bill Vesely	The comment is correct in suggesting that cohort effects can be important. In response to this comment, Appendix M has been added, examining cohort effects on the MSPI in specific examples. These examples suggest that in practice, the cohort effect tends to be insignificant when basic events in a cohort are changed by a factor of less than two, but can be significant when the scale factor is greater than two. Since performance changes of this magnitude are observed in practice, the cohort effect could alter the MSPI value in the neighborhood of the GREEN/WHITE threshold. It is also illustrated in Appendix M that in simple examples, the second-order terms would significantly improve the MSPI approximation. Once the correction coefficients are determined, no additional effort is involved in implementation. However, some effort would be required to obtain the higher-order coefficients from model runs, and many combinations of event types would need to be screened for their significance. Accordingly, for now, the recommendation is to proceed with the MSPI the way it is currently formulated (based on the linear approximation), recognizing that it is an approximation, and that in the neighborhood of the GREEN/WHITE threshold, the change in risk is very small. If the false negative rate on WHITE indications turns out to be significant because of the cohort effect, the second-order correction could be considered for implementation.	Appendix M
13.	The MSPI will increase in its inertia if Bayesian updating is used in successive evaluations of the MSPI. The Bayesian updating needs to be restarted, or other methods applied to keep the inertia in the MSPI from increasing.	Bill Vesely	The MSPI is formulated in such a way that the Bayesian updating is restarted. This approach has the opposite problem; it eschews potentially useful evidence from prior but still recent periods. Future development of mixture priors could help address this (see references in Appendix D).	Appendix D
14.	The MSPI formula is quite complex looking with all the notations included. It would enhance understanding to give simpler alternative presentations. For example, the MSPI for a CDF change due to equipment unavailability changes measured as absolute changes is simply expressed as $$\Delta CDF \cong \sum_j BA_j \bullet \Delta UA_j$$ where BA_j is the Birnbaum importance for equipment j unavailability.....	Bill Vesely	This suggestion has merit. However, for now, this has not been done; significant effort has been invested instead in working with the notation and formulation developed by industry and discussed in a series of workshops over the last several years.	N/A

	Comment	Source	Response	Reference
15.	It is important to calculate uncertainty bounds associated with an MSPI calculated value. Uncertainties are an inherent and an important part of any risk-informed evaluation.	Bill Vesely	As suggested in the comment, a full treatment of uncertainty is typically essential in the context of a general decision analysis. However, the present decision rule does not explicitly consider expected utility based on uncertainty analysis; instead, it is formulated in terms of the mean value of the index, which is to be evaluated based on the CNIP. The probabilities and consequences of false indications of various types have been reflected to some extent in the formulation of this rule (including "frontstops" and "backstops"), based on pilot applications and workshop discussions. This approach does not address the uncertainties in the Birnbaums, but at least acknowledges the uncertainties in the current performance data. If the calculated Birnbaums were to increase substantially as a result of considering within-cutset correlations (presently not taken into account), then some of the thresholds could change. But unless the formulation of the decision rule were to change to use something other than the mean values, knowledge of the spread in the indices would not add value to the agency's current mean-based decision process.	N/A
16.	The Mitigating System Performance Index is substantially superior to the group of safety system unavailability performance indicators, which it replaces. "The Report on the Independent Verification of the Mitigating Systems Performance Index (MSPI) Results for the Pilot Plants" should be issued, its recommendations should be implemented, and the process for incorporating the MSPI into the Reactor Oversight Process should continue. We have previously provided recommendations on the elements of formulation of the MSPIs. Our recommendations have been incorporated into the current MSPI. The necessary elements required to implement the MSPIs are now fully identified and addressed. We commend the staff and industry for developing the MSPIs. These indicators should be implemented promptly.	ACRS	None required.	N/A

NRC FORM 335 (9-2004) NRCMD 3.7	U.S. NUCLEAR REGULATORY COMMISSION BIBLIOGRAPHIC DATA SHEET *(See instructions on the reverse)*	1. REPORT NUMBER (Assigned by NRC, Add Vol., Supp. Rev., and Addendum Numbers, if any.) NUREG-1816

2. TITLE AND SUBTITLE

Independent Verification of the Mitigating Systems Performance Index (MSPI) Results for the Pilot Plants

3. DATE REPORT PUBLISHED

MONTH	YEAR
February	2005

4. FIN OR GRANT NUMBER

NRC Job Codes J8263 and Y6370/Y6636

5. AUTHORS

D. A. Dube[1], C. L. Atwood[2], S. A. Eide[3], B. B. Mrowca[4], R. W. Youngblood[4], D. P. Zeek[3]

6. TYPE OF REPORT

Technical

7. PERIOD COVERED *(Inclusive Dates)*

8. PERFORMING ORGANIZATION - NAME AND ADDRESS *(If NRC, provide Division, Office or Region, U.S. Nuclear Regulatory Commission, and mailing address; if contractor, provide name and mailing address.)*

[1]Division of Risk Analysis and Applications
Office of Nuclear Regulatory Research
U.S. Nuclear Regulatory Commission
Washington, DC 20555-0001

[2]Statwood Consulting
2905 Covington Road
Silver Spring, MD 20910

[3]Idaho National Laboratory
P.O. Box 1625
Idaho Falls, ID 83415

[4]ISL, Inc.
11140 Rockville Pike
Rockville, MD 20852

9. SPONSORING ORGANIZATION - NAME AND ADDRESS *(If NRC, type "Same as above"; if contractor, provide NRC Division, Office or Region, U.S. Nuclear Commission, and mailing address.)*

Division of Risk Analysis and Applications
Office of Nuclear Regulatory Research
U.S. Nuclear Regulatory Commission
Washington, DC 20555-0001

10. SUPPLEMENTARY NOTES
D. A. Dube, NRC Project Manager

11. ABSTRACT *(200 words or less)*

In its Reactor Oversight Process (ROP), the U.S. Nuclear Regulatory Commission (NRC) currently uses performance indicators to quantify safety system unavailability (SSU) for four important nuclear power plant systems. Over time, the NRC staff has identified a number of concerns related to the use of these indicators, including the use of short-term unavailability to approximate unreliability, the use of generic performance thresholds irrespective of variations in risk significance, and potential double-counting as a result of support system failures cascading onto front line systems. Moreover, the way the SSU indicators currently measure unavailability is inconsistent with the definition in the NRC's Maintenance Rule, as well as the indicators promulgated by the World Association of Nuclear Operators and the Institute of Nuclear Power Operations.

This report describes the background, technical issues, and pilot project leading to the development of a more risk-informed performance indicator, known as the Mitigating Systems Performance Index (MSPI). The MSPI addresses most of the concerns related to the use of the current indicators. The NRC staff extensively tested and improved the MSPI methodology during a 12-month pilot plant application phase that involved 20 nuclear power plant units of varying design. The staff also evaluated technical issues related to the new indicator's sensitivity to probabilistic risk assessment (PRA) modeling detail. In addition, the staff compared the MSPI results to the existing indicators, as well as findings from the significance determination process (SDP). The analysis indicates that the MSPI appears to consistently provide a better measure of integrated system performance than the current SSU performance indicators.

12. KEY WORDS/DESCRIPTORS *(List words or phrases that will assist researchers in locating the report.)*

Mitigating Systems Performance Index
Risk-Informed Performance Indicators
Unavailability Performance Indicators
Unreliability Performance Indicators
Plant-Specific Performance Thresholds

13. AVAILABILITY STATEMENT
unlimited

14. SECURITY CLASSIFICATION

(This Page)
unclassified

(This Report)
unclassified

15. NUMBER OF PAGES

16. PRICE